Al-Based Energetic Nanomaterials

Nanotechnologies for Energy Recovery Set

coordinated by
Pascal Maigné

Volume 2

Al-Based Energetic Nanomaterials

Design, Manufacturing, Properties and Applications

Carole Rossi

iSTE

WILEY

First published 2015 in Great Britain and the United States by ISTE Ltd and John Wiley & Sons, Inc.

ISTE Ltd
27-37 St George's Road
London SW19 4EU
UK

www.iste.co.uk

John Wiley & Sons, Inc.
111 River Street
Hoboken, NJ 07030
USA

www.wiley.com

Library of Congress Control Number: 2015936237

British Library Cataloguing-in-Publication Data
A CIP record for this book is available from the British Library
ISBN 978-1-84821-717-1

Contents

Introduction

Over the past two decades, the rapid development of nanochemistry and nanotechnology has allowed the synthesis of various materials and oxides in the form of nanopowders, making it possible to produce new energetic compositions and nanomaterials. Thermite mixtures, intermetallic reactants and metal fuels nanomaterials, often termed as nanoenergetic material, have been widely studied for pyrotechnic applications at large, as a component of traditional gas generating material or more recently as new energetic compounds. The main line driving most of the works in nanoenergetic materials was to enhance the surface area and maximize the intimacy between metal-based reactive components to increase the reaction rate and decrease the ignition delay, while improving safety [BAD 08, DLO 06, DRE 09, ROS 07]. Recently, new insights into the atomic scale description of interfacial regions have provided alternative ways to control the nanomaterial thermal properties [HEM 13, KWO 13]. Advantages of these new metal-based energetic materials, including the addition of further ingredients into the overall propellant and explosive formulations, make it possible to reach not only high energy density, reduced impact sensitivity and high combustion temperature, but also introduce the possibility of producing a wide range of gases upon reaction. These new categories of nanoenergetic materials , also called reactive nanomaterials, should lead to major breakthroughs in pyrotechnics, explosive and propulsion-related materials as well as in small-size integrated pyrotechnic devices. Along this line, recent advances in the integration of nanoenergetic materials into microelectromechanical systems (MEMS) inaugurate the development of "nanoenergetics-on-a-chip" devices, opening up several potential applications in miniaturized pyrotechnical systems as propulsion systems [APP 09, CHU 12, ROS 02], micro ignition and rapid initiation [CHU 10a, ZHA 13, WAN 12, ZHA 08, MOR 10,

ZHO 11, ZHU 11, MOR 11, STA 11d, QIU 12, YAN 14, MOR 13, TAT 13, ZHU 13, LEE 09, BAE 10, HOS 07].

Several other identified applications have also emerged, boosted by the generation of new primers, explosive and propellant additives [STA 10, REE 12, WAN 13], and new materials processing [LEE 09, BAE 10, HOS 07]. On the side, novel "exotic" applications for thermite mixture came up, such as MEMS energy sources [ROS 07], pressure-mediated molecular delivery [ROD 09, KOR 12], material synthesis [RAB 07, KIM 06, MCD 10], biological agent inactivation [SUL 13, GRI 12, CLA 10], hydrogen production [FAN 07, DUP 11] and nanochargers for energy storage [PAN 09b].

This book has a bottom-up structure, from nanomaterials synthesis to the application fields. Starting from aluminum nanoparticles synthesis for fuel application, it proposes a detailed state of the art of the different methods of preparation of aluminum-based reactive nanomaterials. It describes the techniques developed for their characterization and, when available from publications, a description of the fundamental mechanisms responsible for their ignition and combustion. This book also presents the possibilities and limitations of different nanoenergetic materials and related structures, as well as the analysis of their chemical and thermal properties. The whole is rounded off with a look at the performances of reactive materials in terms of heat of reaction and reactivity mainly characterized as the self-sustained combustion velocity. The book ends with a description of current nanoenergetic materials applications underlying the promising integration of aluminum-based reactive nanomaterial into microelectromechanical systems.

We also tried to bring our expertise and experience concerning the application of technologies for the realization of new advanced aluminum-based nanoenergetic materials. After two decades of research, excellent review papers that comprehensively discuss nanoenergetic materials, especially concerning aluminum-based reactive materials, with numerous citations therein are referenced for the benefit of this book. We encourage the readers to consult them [DRE 09, ROS 07, ROG 10, ROG 08, ROS 14, ROS 08, ADA 15].

Acknowledgements

First, I thank my colleague Dr. Alain Estève, CNRS researcher, who provided insight and expertise that greatly assisted the research and for his comments that greatly improved the book. I also thank all my phD students and post-docs who conducted all the technical stuff. The list is long and I prefer to stress the attention to Dr. Gustavo Ardila-Rodriguez, Dr. Marine Pétrantoni, Dr. Guillaume Taton, Dr. Jean Marie Ducéré, Théo Calais, Ludovic Glavier and Vincent Baijot. I would like to express my gratitude to Dr. Daniel Estève, Prof. Mehdi Djafari-Rouhani and Véronique Conédéra for helping me in my research. Last but not least: I apologize to all of those who have been with me since 1997 and whose names I have failed to mention.

1

Nanosized Aluminum as Metal Fuel

The replacement of micrometer-size metal fuel such as aluminum (Al) or boron (B) powders in solid propellants, explosives and pyrotechnics with their nanometer-size counterpart (Nanosized A1) has become a common trend in the design of new types of propellants and solid fuel in recent decades. The utilization of nanosized particles is shown to: (1) shorten the initiation; (2) shorten burn times to increase the completeness of the combustion and therefore, to improve specific impulse; (3) enhance heat-transfer rates from higher specific surface area and; (4) enable new fuel/propellants mixture with desirable physical and energetic properties. Moreover, the nanoscale control of their synthesis together with their tuned properties authorizes new perspectives for their use, for instance, as solid fuels in automotive engines [KLE 05].

Different techniques have been developed for synthesizing nanopowders of different natures, sizes and shapes, but the emphasis is put on nanopowders of aluminum which are mostly used in practice to dope propellants, explosives and pyrotechnics. It offers a reasonably high-energetic density source and is also largely available in the Earth's crust for the benefit of mass production capability [STA 10, REE 12, WAN 13, DUB 07]. The oxidation of aluminum to alumina (Al_2O_3) releases –31.1 kJ/g [LID 91]. By comparison, CL-20 ($C_6N_{12}H_6O_{12}$) has an enthalpy of combustion of 8 kJ/g [SIM 97]. Boron is also a good choice as an additive since the oxidation of B into B_2O_3 releases –58.9 kJ/g; however, the presence of the low melting oxide on the particle surface and the formation of hydrogen boron oxygen (HBO) intermediate species (HBO, HBO_2) slow the combustion and in consequence, the rate of energy release.

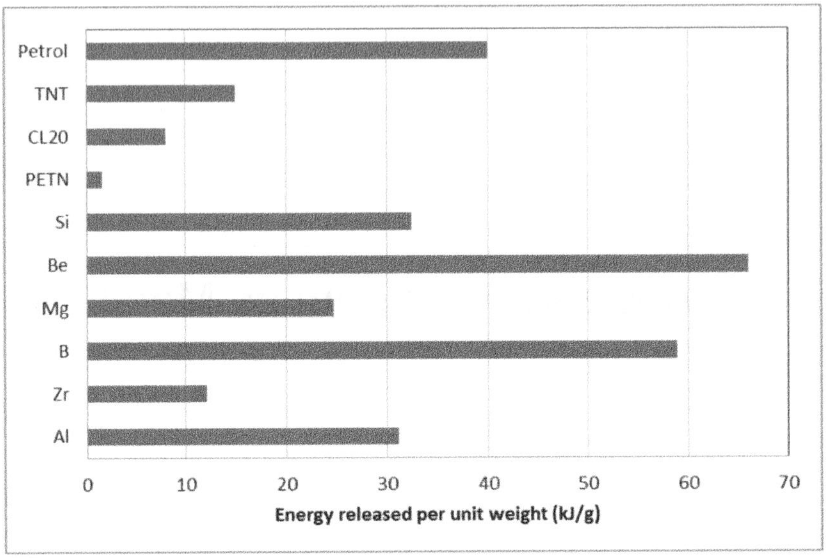

Table 1.1. *Maximum enthalpies of combustion for selected monomolecular energetic material in comparison to a few metal fuels*

1.1. Al nanoparticles manufacturing

The rapid acceleration of research in the area of nanoenergetic materials is mainly connected to the progress made in the manufacturing of Al nanopowders that made it possible to increase and multiply the number of research experiments in laboratories, more than a decade ago. In the following, we discuss the different methods for producing Al metallic nanoparticles that can be classified into three distinct categories: (1) those based on vapor-phase condensation; (2) those based on liquid phase chemistry and to a lesser extent; (3) those based on mechanical methods.

1.1.1. *Vapor-phase condensation methods*

1.1.1.1. *Electrical explosion and vaporization wire*

Most of the studies describing Al nanoparticles or including them into composite energetic materials use Al nanopowders synthesized by electrical explosion wire (EEW) process under diverse atmospheres. The method, which has its roots in the work of Narme and Faraday (1774), has been pioneered for metal nanoparticles fabrication by Russian scientists starting in the late 1980s [DOL 89] and continues to be developed around the world since then [SED 08, IVA 03, JIA 98, KWO 01, SAR 07]. The electrical explosion is accompanied by shock-wave generation and rapid heating of the metal to a temperature of 10^4 °C at a rate of more than 10^7 °C/s. The underlying physics of the wire explosion remains the subject of current investigations. However, there is consensus in the fact that an explosion occurs forming a plasma. This plasma is spatially restricted by a very high field created by the pulse. When the metal vapor pressure exceeds the cohesive force of the metal, there is an interruption in current flow, causing the plasma to generate clusters of metal that are projected at supersonic speeds in the environment. EEW technology is used to produce nanopowders of Al, Ti, Zr, Mg and other metals with a particle size of 40–100 nm and a specific surface area of 10–50 m^2/g. The method is employed on a large scale with a production capacity of a few hundreds of grams per hour with a rate depending on the metal type.

Even if the process is performed in an inert atmosphere (e.g. He, Ar or Xe), pure aluminum being pyrophoric, the aluminum particles are spontaneously passivated with a thin alumina layer. This natural thin alumina layer that is formed spontaneously at low temperature is amorphous with a thickness ranging from 0.5 to 4 nm. Most of the experimental data on commercial Al particles give a thickness ranging from 2 to 3 nm. A way to control the thickness of the particle oxide layer is to passivate the nanoparticles with a controlled protective oxide (see Figure 1.1) just after the nanoparticle formation to effectively hinder further oxidation during their storage. Practically, the as-grown aluminum nanoparticle will be very sensitive to whatever oxidizing atmosphere leading to different alumina layers in nature and thickness, such as the formation of hydroxide. The control of this passivation phase is commonly accomplished as a separate processing step, in which the chamber filled by inert gas for powder production is evacuated and refilled with an oxidizing gas mixture. Typically, a dry oxidizing atmosphere with low partial pressure of oxygen (e.g. 0.01% of the total pressure) is sufficient to control the passivation. Alex® is a leading manufacturer of powders produced by the EEW technique [SAR 07, TEP 00]. Characteristic transmission electron microscope (TEM) images of ALEX® aluminum nanopowders obtained by wire explosion process are shown in Figure 1.2.

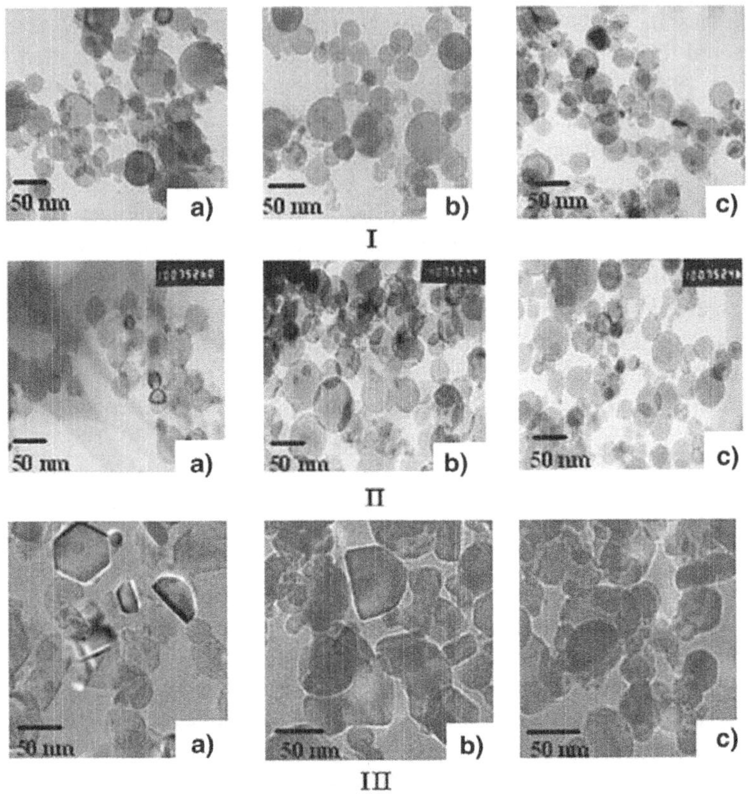

Figure 1.1. *Transmission electronic microscopy images of aluminum nanoparticles produced in different atmospheres: (I) helium, (II) argon and (III) nitrogen and for different oxygen pressures a) 0.025 MPa, b) 0.05 MPa and c) 0.1 MPa [SAR 07] (Copyright 2007 Elsevier)*

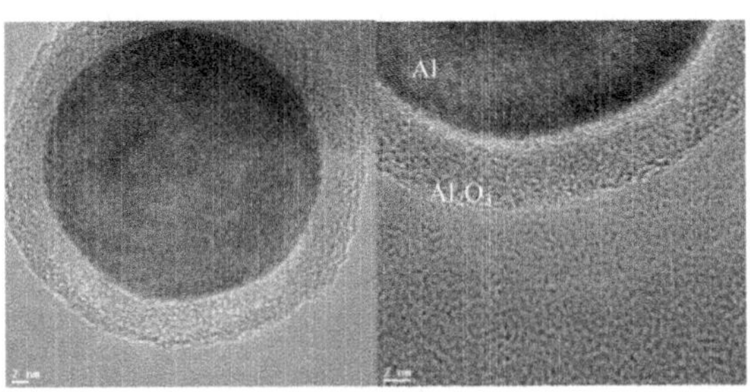

Figure 1.2. *Transmission electronic microscopy images of ALEX® aluminum nanoparticles. Pure aluminum core coated with a 3–4 nm alumina shell is observed*

Aluminum nanopowders can also be produced from condensed aluminum vapor generated when a thin aluminum wire is vaporized by a strong electric current passing through it. After condensation, the particles are collected on the walls of the explosion vessel. Nanoparticles of many other metals and alloys can also be obtained such as Cu, Ni, Fe, Cu/Zn, TiO_2, TiN, Fe_2O_3, etc. [TEP 00, WAN 01a, KWO 05, WAN 01b, NAK 98]. The effects of pressure, gas environment, electric pulse characteristics and other experimental parameters have been widely studied [JIA 98, IVA 03, KWO 01, SAR 07]. It was demonstrated that a higher pressure results in the formation of coarser particles. It was also reported that an increased pressure of the inert gas results in an increased yield of aluminum nanoparticles. This synthesis procedure enables to produce the Al nanopowders in different environments. By varying the composition and concentration of the atmosphere, the composition of the passivation layer can be tuned. For example, aluminum particles coated with a thin layer of AlN, $Al(OH)^3$ or n-Al_4C_3 can be obtained by performing vaporization or electrical explosion in a nitrogen–argon gas mixture or water or decane. Other passivation layers have also been considered as fluoropolymers, stearic and oleic acids, and aluminum diboride [GRO 06b, KWO 03a]. In Figure 1.3, we see the two-layer coating of organically passivated Al particles. A detailed section on passivation layer is proposed later on in the book.

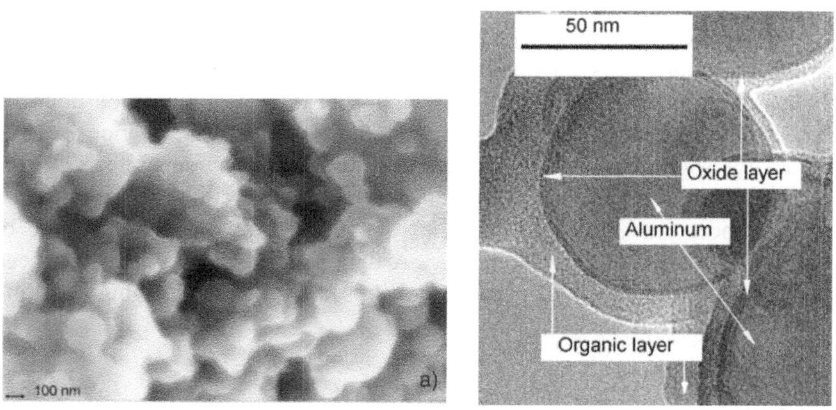

Figure 1.3. *Scanning electronic microscopy images of aluminum nanoparticles passivated by steric acid [GRO 06b] (Copyright 2006 Wiley)*

Electro-exploded wires technology presents the advantages of being simple, efficient and well-adapted to producing large quantities of nanoparticles. The sizes of the obtained particles typically vary over a fairly broad range from 40 to 100 nm. The control of the size and shape of the particles are operated through the adjustment of the energy passing through the wire for its vaporization and the use of a suitable medium [WAN 01b, NAK 98].

1.1.1.2. *Other techniques*

A couple of other techniques for the production of Al nanoparticles have been proposed and used. The evaporation of bulk aluminum samples or aerosolized micron-sized powders followed by controlled vapor condensation has been discussed in the literature [GRA 76, SCH 06, PIV 06, CHA 06]. A crucible containing bulk aluminum is heated until vaporization in a flowing inert gas environment. The heating can be performed through several different protocols: radiative heaters, induction heaters, lasers, electric arcs or special high temperature furnaces. The pressure and nature of gas during condensation contribute to the determination the particle characteristics: a low pressure (less than 1 kPa) inert gas results in the formation of nanoparticles, while high pressures result in increased particle sizes. Additionally, condensation of metal vapors in lighter inert gases (e.g. He, Ar or Xe) is shown to produce finer particles.

Cryogenic melting of metals [BRE 08] is another effective method for producing metal nanopowders. In this case, metal particles are formed as a result of spontaneous condensation of metal vapor in a cryogenic liquid. Metal vapor is produced by rapid induction heating of a metal rod due to a powerful high generator. This heating method produces high vapor pressure in a very short period. The cryogenic liquid is continuously fed into the reactor, and nanoparticles are formed by rapid condensation of the saturated metal vapor. The low temperature of the medium provides a high rate of particle formation and rapid cooling. A diameter less than 70 nm has been demonstrated. The method is particularly used for metals with high melting points.

1.1.2. *Wet chemistry*

Wet chemistry techniques are attractive for the commercial synthesis of aluminum nanopowders because of their setting-up simplicity, the inherent safety of handling the reactive powder under liquid and the ability to functionalize the particle surface during the manufacturing process. In

[HIG 01], Al nanopowders were prepared by decomposing alane-adducts in organic solvents under an inert atmosphere. Effective adduct species were reported to include trialkylamines, tetramethylethylene-diamine, dioxane and other aromatic amines and ethers. Highly uniform particles were obtained with particle size selectable in the range of about 65–500 nm by adjusting the catalyst concentration and by varying the concentration of the adduct species. As is typical for all reported wet chemistry techniques, the methodology is based on careful and slow mixing of measured amounts of the starting solutions followed by continuous stirring and drying of the product. Such a methodology is not well suited for mass production and substantial modifications are necessary to obtain practical quantities of the desired reactive nanopowders. As a result, to date, there is no commercial mass production of aluminum nanopowders produced by wet chemistry techniques.

1.1.3. *Mechanical methods*

Milling processes can be an alternative to gas condensation methods or wire electro-explosion processes to prepare highly reactive particles. In [AND 13], Andre *et al.* proposed mechanical milling approach for the synthesis of reactive aluminum nanoparticles. Milling techniques are currently explored for the manufacturing of nanostructured powders with enhanced properties in comparison to micro-sized particles for different types of materials [PAT 12]. These mechanical methods are well adapted for the generation of aluminum nanoparticles due to the high ductility of aluminum material. The introduction of oxygen during milling enables us to obtain powders with high specific areas: specific surface area obtained is for the best around 20 m^2/g, but the nanoparticle shape is not spherical. Interestingly, the core aluminum is poly-crystalline (see Figure 1.4) and the amorphous alumina shell thickness is superior to the one obtained with previously presented methods. Specifically, Andre *et al.* have measured a thickness of 4.5 +/–0.5 nm for Al nanopowders synthesized using a milling technique in the following conditions. Powder from AlfaAesar[1] (purity of 99.8%) was milled using a high-energy planetary ball milling equipment. The rotation speeds of vials and plateau were 800 rpm (round per minute) and 400 rpm, respectively, in opposite directions for a total milling time of 16 h. A

1 https://www.alfa.com/.

controlled quantity of air (71 cm^3) was added in order to monitor the thin alumina layer formation.

Figure 1.4. *Transmission electronic microscopy images of aluminum nanoparticles obtained by high-energy ball milling: a), b) aggregation of particles; c) zoom on the aluminum core poly-crystallinity; d) zoom on the amorphous alumina layer [AND 13] (Copyright 2013 Elsevier)*

One of the main differences with nanoparticles obtained by vapor-phase condensation is the presence of alumina at the grain boundaries.

1.2. Example of Al nanoparticles passivation technique

The surface passivation of aluminum particles is of crucial importance for safety and handling purposes as well as for its impact on thermal properties and energetic performance. As previously mentioned, pure aluminum being pyrophoric, the aluminum particles are spontaneously passivated with a thin amorphous alumina: from 0.5 to 4 nm in thickness depending on the time of exposure with oxygen (mainly linked to the storage time). Current research on Al nanopowder is focused on preparing high-quality and well-passivated powders [FOL 05b, JOU 05b] following different strategies detailed in the next sections. The goal for all the studies is to develop passivation coatings that can protect particles from oxidation when stored in air without affecting the particle burning properties. Furthermore, preventing the formation of further alumina passivation layer increases the purity (i.e. rate of pure aluminum over the entire product) of the powder guaranteeing the reliability of its energy reservoir. Typically, Al particles made from vapor condensation, passivated with a native thin alumina, have purity ranging from 42 to 81% depending on the particle diameter ranging from 20 to 80 nm, considering a passivation shell thickness of 2 nm. With a passivation coating preventing spontaneous oxidation, the metal content in the powder can reach 95–98 wt.% of metal Al and therefore, increase considerably the percentage of pure aluminum per unit of mass. We can distinguish metal-based passivation layers and organic coatings.

1.2.1. *Metallic coating*

For example, surface layers of transition metals were formed on Al nanoparticles to prevent them from oxidation in air [FOL 05, GAO 07]. In [FOL 05], aluminum nanopowder was synthesized by thermal decomposition of an alane solution in the presence of a titanium catalyst under an inert atmosphere. In [GAO 07], the nanopowder was formed upon mixing and drying of aluminum dissolved in NaOH co-mixed with a nickel salt solution. In both cases, aluminum nanoparticles served as a reducing agent for the transition metal complexes so that reduced metal films were produced on the aluminum surface.

1.2.2. *Organic coating*

In a different passivation approach, aluminum nanopowders coated with non-metallic self-assembled monolayers (SAMs) were prepared in [JOU 05a, JOU 06]. Nanoscale Al particles were produced in solution by the catalytic decomposition of $H_3Al.NMe_3$ or $H_3Al.N(Me)Pyr$ by $Ti(O^iPr)_4$ (see [JOU 05b] for details) and coated *in situ* using a perfluoroalkyl carboxylic acid SAM. The Al particles were coated directly in solution and therefore, were not exposed to oxygen. It was observed that the SAM coating passivated the aluminum preventing the oxidation of the particles once exposed to air. Contrary to alumina-coated nanoparticles, it makes the nanopowder soluble in polar organic solvents such as diethyl ether [JOU 05b, JOU 06]. The obtained composite particles (A1 coated with organic) are shown in Figure 1.5. The protective layer is clearly fairly thick resulting in the overall reduction of the energy density of such materials.

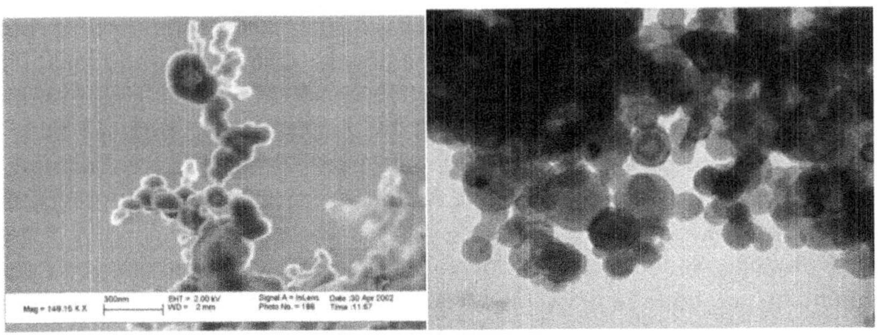

Figure 1.5. *Scanning electronic microscopy images of the Al/$C_8F_{17}COOH$ composite at 148 000 magnification and Al/$C_{13}F_{27}COOH$ composite at 200 000 magnification [JOU 05b] (Copyright 2005 American Chemical Society)*

New approaches to passivating the surface of Al nanoparticles continue to be developed and substantial progress is expected in this area in the near future. For example, encapsulation of aluminum nanopowders in polystyrene was recently described in [ZHA 07] and was shown to be effective in preventing aluminum oxidation.

In [PAR 06], Rai *et al.* studied the properties of carbon as a passivating layer, by using either a laser ablation method or a direct current (DC) arc to create, and coat aluminum nanoparticles in a continuous manner. The carbon coating was created by injecting ethylene (C_2H_4) into the plasma created by the laser. As a result, the aluminum nanoparticles were coated with a carbon layer of thickness around 1–3 nm. Below 700°C, the coating was stable. However, the carbon coating oxidized above 800°C.

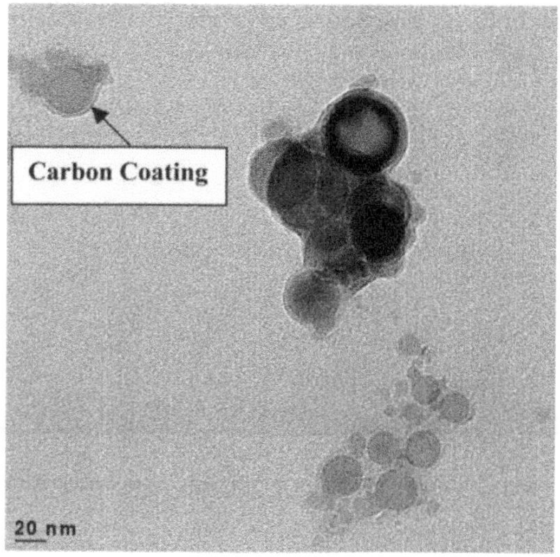

Figure 1.6. *Transmission electronic microscopy images for coated aluminum nanoparticles obtained by DC arc method [PAR 06] (Copyright 2006 Springer)*

1.3. Characterization of Al nanoparticles properties

The key properties of aluminum nanopowders to be known before any use in energetic mixtures or for energetic applications include particle size, size distribution, particle morphology, chemical composition, nature and thickness of the passivation layers and the percentage of pure metal. These data are of primary importance in determining the reactivity and ignition temperature of the Al powder. Various methods exist to unravel nanoparticle structures and measure powder size distribution. This section presents the different techniques and methods found in the literature to picture aluminum nanopowder structures and to subsequently study their thermal response in the context of energetic materials.

The size and particle morphology, the size distribution can be obtained by electron microscopy as, for example, scanning electron microscopy (SEM) and TEM. They are common tools and practices to determine the particle shape, size and give information about the particle surface morphology (see Figures 1.6 and 1.7). In order to be representative, such measurements must consider a large number of particles and different locations of the sample. The determination of particle size distribution is then carried out by image analysis. In general, a powder is considered monodisperse when the diameters of the individual particles fall within 10% from each other.

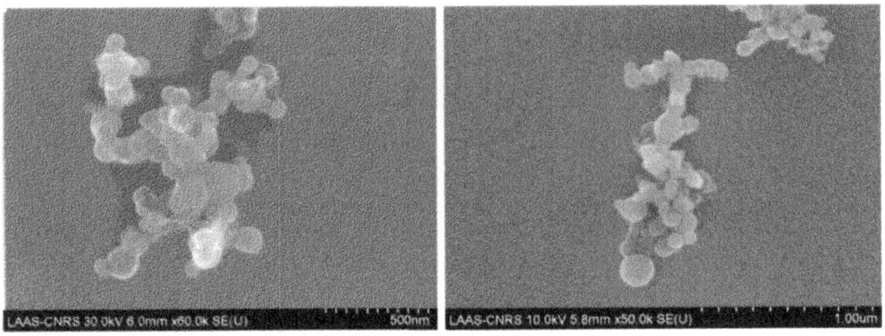

Figure 1.7. *Scanning electronic microscopy images of nanometer aluminum particles*

High resolution transmission electron microscopy (HR-TEM) is also successfully used to characterize the oxide coatings present on aluminum particles exposed to different environments. As an illustration, we can go back to Figure 1.2 showing the Al nanoparticle core and its surface passivation layer, many other examples can be found in published papers [GRO 06a, RUF 07, RAM 05, SEV 12]. Crystallinity, microstructure and homogeneity of the oxide layers are inherently difficult to measure due to their overall amorphous nature.

1.3.1. *Light scattering methods*

Particle sizes can also be quantified using laser light scattering (LS) and several commercial instruments utilizing this technique are available. Low-angle light scattering (LALS) is the only LS technique that actually measures a scattered light intensity that is due to the interaction between a single particle and an incident light. The scattered light intensity is thus directly proportional to particle diameter and weight. The scattering behavior is also dependent on the light wavelength. These instruments are generally designed for size characterization of micron-sized powders or their aggregates. However, they are also capable of measuring smaller sizes, typically down to the 50–100 nm range due to specific algorithms for processing the scattered laser emission spectra. The main advantage of this type of measurement is that a powder with relatively broad particle size distribution can be characterized. However, for accurate measurements, the optical properties of the material surfaces need to be known. Such properties are not well established for many materials, especially for the nanosized particles for which, as noted above, the material properties are expected to substantially differ from those of the respective bulk materials.

Another group of laser LS devices is the dynamic LS or photon correlation spectroscopy. A laser light is focused into a suspension. It determines the particle size by measuring the rate of fluctuations in laser light intensity scattered by particles which have been dispersed in a liquid [BAR 84, PEC 85]. The Brownian motion of particles in suspension causes laser light to be scattered at different intensities. Analysis of these intensity fluctuations gives the velocity of the Brownian motion and hence, the particle size using the Stokes–Einstein relationship. The measurement needs to be processed assuming the viscosity of the fluid, a specific shape of the size distribution function, so that generally the mean particle size and the width of the particle size distribution can be quantified. Dynamic LS technology offers many advantages such as accurate repeatable particle size analysis in short time and reduced sample size.

1.3.2. *Gas adsorption method: specific surface measurement, BET diameter*

Besides electron microscopy, one common characterization technique for the determination of particle size is based on the volumetric adsorption of a gas on a sample of particles that provide a measurement of the specific surface of these particles. Surface areas are calculated using so-called Brauner, Emmett and Teller (BET) theory. Usually, the surface area measurements are determined by nitrogen adsorption using a gas adsorption analyzer. The "BET diameter" assumes that the measured specific surface of a powder is equivalent to that produced by monodisperse spherical particles, which is not necessarily the case in nanopowders [MOO 07, PRE 05]. This technique and associated theory are well known, widely used and a few commercial devices are available. The BET method is shown to provide more accurate experimental data on ultra-fine particle samples.

1.3.3. *Thermal analysis: purity or aluminum content percentage and oxide thickness*

Thermal analysis techniques are widely used for the characterization of energetic materials, especially differential scanning calorimetry (DSC) and thermogravimetry analysis (TGA). In this latter technique, a small amount of powder, typically 5–10 mg, is placed in a crucible and is heated at low heating rate (5–40°C/min) under flowing air or mixture of Ar and oxygen gas. The sample is weighted before and after the analysis. For DSC, after the first heating cycle, the sample is cooled down to room temperature and then heated again at the same heating rate. This second analysis is used to correct the baseline. It is assumed that the bulk heat capacity of the sample does not change between the first and the second heating run. Note that thermal analysis relies on assumptions about the density, structure and homogeneity of the natural passivation layer. The sample has to be well characterized before thermal analyzing with respect to its detailed chemical composition, its particles size, its size distribution, etc., to ensure a good thermal measurement accuracy. In practice, a sufficient number of repeated measurements are required.

Figure 1.8 shows a typical TGA curve of the Al nanopowder oxidation in Ar + O_2 environment in the temperature range of 30–1,000°C at a heating rate of 10°C/min. The active aluminum content can be easily obtained from mass gain measurement in a thermogravimetric analyzer. The mass gain Δm(%) in the TGA under Ar/O_2 mixture (typically, 75/25 mixture) is attributed to oxidation of active aluminum:

$$4Al + 3O_2 \rightarrow 2Al_2O_3$$

The aluminum content, or purity P, can be calculated using the following equation based on the mass balance and the mass ratio of aluminum to oxygen in Al_2O_3:

$$P = \frac{108}{96}\Delta m(\%) \qquad [1.1]$$

P is typically between 41 and 80% for particle size ranging from 20 to 80 nm and considering an oxide shell thickness of 2 nm. Knowing P and the total particle diameter d, the thickness of the oxide layer (t_{oxide}) is calculated by equation [1.2]:

$$t_{oxide} = \frac{d}{2}\sqrt[3]{1 - \frac{P\times\rho_{Al2O3}}{m_{Al2O3}+P\times(\rho_{Al2O3}-\rho_{Al})}} \qquad [1.2]$$

with $\rho_{Al_2O_3}$ (3.05 g.cm^{-3}) and ρ_{Al} (2.7 g.cm^{-3}) being the densities of amorphous aluminum oxide and aluminum at room temperature, respectively. Note that it is assumed that Al_2O_3 is amorphous with a density of 3.05 g.cm^{-3}.

However, it has been reported that the alumina passivation layer can be porous, partially hydrated and certainly non-uniform in thickness [SEV 12, BAR 84]. Recent research using prompt gamma neutron activation analysis and HR-TEM [GAO 07] showed that the amorphous oxide coatings on aluminum nanopowders can also include impurities of hydroxide, trapped water molecules and boron. The oxide thickness can be determined by HR-TEM images combined with particle size distribution measurement.

The combination of multiple characterization tools (DSC, TGA, TEM...) makes it easier to properly estimate the gravimetric fraction of pure Al in the powder, *P*. Data from TGA curves can be also used to determine activation energies [AUM 95], pure or active aluminum content [JOH 07], oxide layer thickness [EIS 04] and particle size.

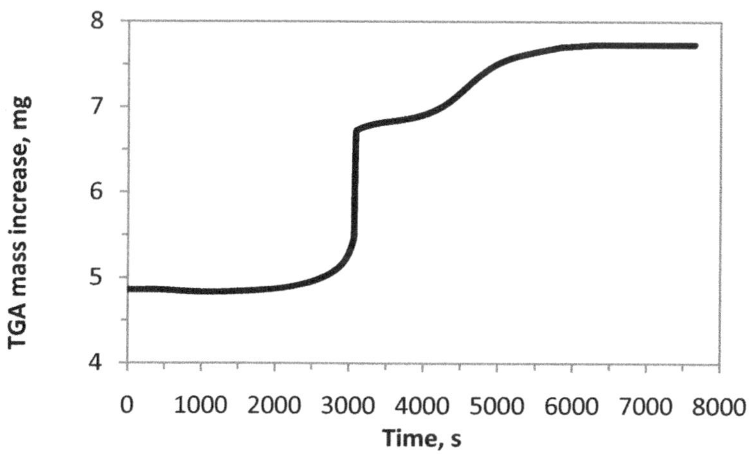

Figure 1.8. *TGA curves of aluminum nanopowder in Ar/O_2. The TGA scans were performed at 10°C/min under 25/75 O_2/Ar atmosphere*

1.3.4. *Chemical analysis*

Chemical analysis using induced coupled plasma (ICP) emission is another tool that enables us to determine the purity of nanopowder as well as the Al/Al_2O_3 ratio. The metallic Al content is derived from the oxygen content by assuming that oxygen is only present in the oxide passivation layer.

1.4. Oxidation of aluminum: basic chemistry and models

In this section, we consider aluminum powder oxidation under low heating rate, e.g. < 40°C/min. It is now well established that oxidation of nanopowders begins at much lower temperatures than their micron size counterparts. The oxidation rate of Al depends on the temperature of the material and can be represented on the typical curve given in Figure 1.9.

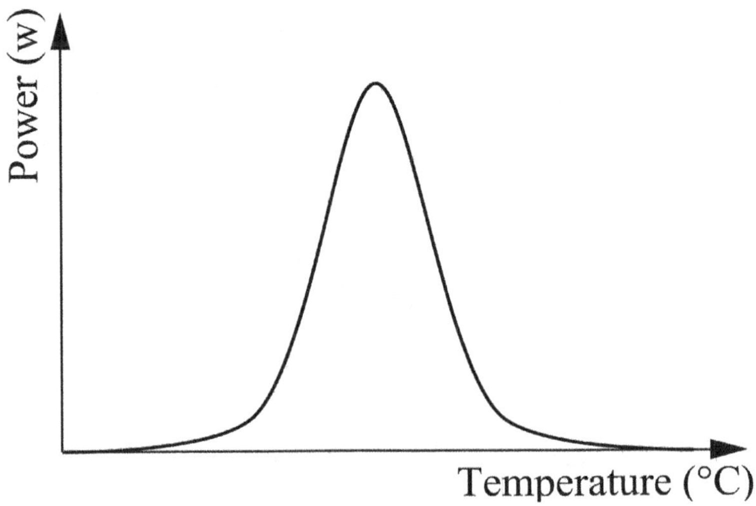

Figure 1.9. *Typical DSC curve of the aluminum oxidation in O_2/Ar oxidizing environment*

The complex oxidation process leading to the formation of the alumina passivation layer is still not well understood but few chemical reactions and phenomenological mechanisms responsible for oxidation have been proposed in the literature to fit the experimental results. They will be presented in the following sections.

1.4.1. *Initial stage of aluminum oxidation from first principles calculations*

The early stage of aluminum oxidation ($n = 0$) has attracted much interest because of its practical importance and also because it was considered as a model system to understand the fundaments of the oxidation of metallic surfaces. On the experimental side, the first comprehensive picture of the aluminum oxidation at early stages and its different regimes was given by Brune *et al.* in the early 1990s with the help of STM experiments [BRU 92, BRU 93]. At low coverage (a few Langmuir) and room temperature, molecular oxygen exhibits dissociative chemisorption. The oxygen atom separation following the dissociation is shown to be enhanced by their hyperthermal motions because of the exothermic nature of the chemical decomposition reaction. This process is accompanied with surface diffusion leading to the formation of oxygen (1×1) patterns, i.e. clusters of adsorbed oxygen atoms on high surface symmetry hollow sites. Increasing the coverage, but still long before the oxygen (1×1) monolayer coverage, oxidation starts to occur at the oxygen cluster sites. This means that a mixing of both aluminum and oxygen species is initiated. Since the thermodynamics of the adsorption and formation of oxygen (1×1) clusters plus proposals of potential interfacial oxide/Al structures have been proposed theoretically by density functional theory calculations [BEN 03, JAC 95, RUB 02, KIE 01, KIE 02, CIA 04], a mechanistic description of these observations has come to an end only very recently [LAN 12, LAN 14]. Indeed, it was trivially expected that oxidation process was driven by oxygen penetration into the aluminum substrate or subsurface [ZHU 03] leaving unclear the kinetics and associated pathway for this penetration as well as conditions for this penetration to occur, i.e. at which oxygen coverage and through which atomistic environment. Also, controversial discussions were oriented on the initial coverage value necessary to observe oxidation nucleation [CIA 04]. Lanthony *et al.* confirmed the Brune's STM observations that the oxygen (1×1) clusters were promoting oxidation. They also demonstrated that oxidation nucleation was due to additional incoming of oxidizing species on the outer surface of the oxygen (1×1) clusters making some aluminum atoms underneath the oxygen cluster to "percolate" through it up to the surface. This complex chemical process has been called "extraction" mechanism and its different stages are illustrated in Figure 1.10 when the surface is exposed to atomic oxygen. Importantly, this process is barrierless indicating that oxidation will take place spontaneously as soon as oxygen clusters are formed and exposed to further gas phase oxidizing species. It is shown that under higher coverage, multiple barrierless extractions can be operated, up to the stripping or oxidation of about two layers of the aluminum surface. This leads to the formation of a native and stable aluminum oxide layer. Continuing the oxidation process necessitates activation and other chemical mechanisms that are still to be determined in order to completely understand passivation layer formation.

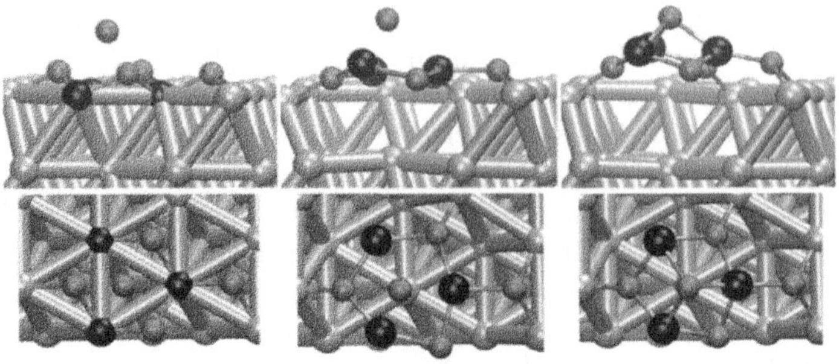

Figure 1.10. *DFT structures of Al "extraction" mechanism: side views on three top images, top views at the bottom. Extracted Al atoms are the biggest in black, O atoms are the smallest in dark gray, gray spheres are aluminum atoms*

1.4.2. *Thermodynamic modeling of Al oxidation under low heating rate*

At the experimental level, the oxidation thermodynamics are explored by combining several techniques. As already mentioned, the oxidation steps are characterized using TGA analysis in oxygen/argon (25/75) environment from room temperature to 1,300°C. The TGA curve gives the mass gain resulting from the oxidation process, and simultaneous DSC measurements can be added to quantify the corresponding exothermic effects as well as to detect phase transitions. The structure and nature of the powders can also be checked before and after thermal treatments by X-ray diffraction or Raman analysis. These experimental results [JOH 07, MEN 98b, JON 00] are generally consistent between each other and show that for particles of 100 nm and finer, the oxide is mostly formed below the bulk aluminum melting point, between 450 and 605°C and with a higher degree of reaction.

The controversy arises during the interpretation of the results since the real mechanism responsible for the oxidation rate of aluminum nanopowders is still not well understood. Some suggest that the Al core expands and breaks the shell ejecting small molten clusters of Al at high velocity [UME 07]. Others suggest that aluminum from the core and oxygen from the outer surface atoms diffuse and meet to react within the passivation layer. This is discussed in the following and will be reviewed in more detail in Chapter 5.

1.4.2.1. *Fixed ignition temperature model*

The simplest and first ignition model found in the literature is based on a number of experimental reports showing that large aluminum particles

(> 10 μm in diameter) ignite close to the melting point of Al_2O_3 (i.e. 2,042°C) [BOI 02, BRO 95]. Hence, it was suggested that the oxide film that is always present on the aluminum surface prevents ignition until it starts melting. This melting results in the formation of isolated oxide islands in replacement of the previous continuous coating film. The surface of aluminum is then exposed to the oxidizer and the particle ignites. Therefore, the description of aluminum ignition can be envisaged at a constant temperature. It has been recognized that a better agreement with experiments can be achieved in most cases if the ignition temperature is selected somewhat below the Al_2O_3 melting point in the sense that a thin and constrained protective oxide has its melting point lower than the bulk value. For experiments in which large aluminum particles (over micron size) are used, this model produced quite satisfactory results. However, this approach is inadequate for finer aluminum particles.

1.4.2.2. *Stress in the oxide layer model*

This model introduced by Rozenband *et al.* [ROZ 92] suggests that the protective oxide coating breaks up due to the increasing mechanical stresses during heating and core Al melting. Then the aluminum surface becomes exposed to the oxidizer and the particle ignites. This hypothesis was largely discussed: the mechanical properties of the heated alumina are poorly known. The most significant stress in the oxide film is obtained just after the melting of the Al shell composed of pure aluminum, i.e. immediately above 660°C. At elevated temperatures, e.g. above 1,100°C, the deformations of Al_2O_3 result in grain boundary sliding causing a brittle to ductile transition. Therefore, the stresses in the oxide film can be relaxed avoiding the rupture of the oxide film. This model is contested to adequately describe ignition of aluminum powders observed at low temperatures (around 500–600°C).

1.4.2.3. *Oxidation growth modeling*

The most conventional approach to modeling the oxidation growth kinetics is to consider a linear-parabolic regime. This approach considers that the initial growth regime is dominated by chemical reactions with the oxidizing species of the gaseous environment. The overall oxidation can be written as [FED 03].

$$\frac{dt_{oxide}}{dt} = K(t_{oxide})^{-n} C_{ox}^{m} exp\left(-\frac{E}{RT}\right) \qquad [1.3]$$

where t_{oxide} is the thickness of the oxide film, t is time, C_{ox}^{m} is the oxidizer concentration proportional to the oxidizing pressure near the particle surface and m is the order of the reaction in terms of the oxidizer. R is the universal gas constant, and E and K are the activation energy and pre-exponent, respectively. The exponent n determines the dependence of the oxidation rate on the oxide film thickness. $n = 0$ for linear oxidation law (suitable for the first oxidation step) and $n = 1$ for parabolic law of oxidation when diffusion of oxidizing species through the oxide already formed is the limiting step for the oxidation kinetics.

In another formulation, the general kinetic equation of oxidation can be represented in dimensionless form as:

$$v_{oxidation} = k_o \eta^{-n} exp\left(-\frac{E}{RT}\right) \ exp(-k_1 \eta) \qquad [1.4]$$

where $v_{oxidation}$ is the rate of the reaction, η is the degree of oxidation determined from TGA mass changes, k_0 is the pre-exponential factor, E is an activation energy accounting for the oxidation chemical reaction, R is the universal gas constant, T is the reaction temperature, and n and k_1 are, respectively, exponent and coefficient constants. E and k are determined from independent sources. Typically, these values are extracted from experimental aluminum oxidation studies. Activation energy usually does not depend on particle size and is ~155 kJ/mol in average for aluminum oxidation. The activation energies for heterogeneous aluminum oxidation found in the literature range from 70 to 420 kJ/mol depending on the temperature range. For example, in [TRU 06], Trunov *et al.* reported activation energy of 460 kJ/mol for 950–1,300°C range and 71 kJ/mol determined for 1,600–2,000°C temperature range. k_o is found to be 2 to 5 times higher for nanopowder than for larger size powders. Note that a small change in the activation energy greatly affects the oxidation kinetics described by an exponential function. Therefore, an adequate selection of the values of E and k is critical for an accurate description of the oxidation processes of aluminum.

Experimental works showed that at the early stage of oxidation ($\eta = 0$), the reaction rate does not depend on the degree of oxidation and the process is similar to monomolecular reactions. The Oxidation process follows a linear law up to a degree of oxidation of 0.1–0.3 and leads to the rapid oxidation of the aluminum surface into a thin amorphous oxide layer. In a second stage, the oxidation process slows down and the oxidation follows a logarithmic law due to the formation of an oxide layer through which the oxidizing species must migrate toward the interface with pure aluminum to continue the oxidation process.

Other choices have also been discussed in the literature based on the experimental data collected for different oxidation temperature ranges [KUB 62]. From this approach, the heat release due to the oxidation is estimated and compared to external particle heating or heat losses. Ignition can, therefore, be predicted by an adequate particle heat transfer model including the enthalpy of oxidation. However, the analysis of the literature shows great discrepancies between the values reported and used by different researchers.

1.4.2.4. *Diffusion based on phase transformation processes*

Recent research on aluminum particle ignition has proposed that the oxidation is controlled by non-constant diffusion processes of oxygen and aluminum through the surface oxide coating. Importantly, it is assumed that the diffusion is critically affected by polymorphic phase changes occurring during heating. Trunov et *al.* [TRU 06] suggested that the oxidation under low heating rates (< 40°C/min) is caused by a sequence of polymorphic phase transition occurring in the oxide layer: amorphous $Al_2O_3 \rightarrow \gamma \rightarrow \delta \rightarrow \theta \rightarrow \alpha$ Al_2O_3

The densities of these alumina polymorphs are different from each other, as shown in Table 1.2.

	Density, g/cm^3
Amorphous Al_2O_3	3–3.1
γ-Al_2O_3	3.6–3.67
α-Al_2O_3	3.99

Table 1.2. *Density of different alumina polymorphs*

The overall phase transformation process can be explained as follows. After the first oxidation step (formation of a thin amorphous oxide), the thickness of the natural amorphous alumina layer increases slowly. Around 550°C, this layer exceeds the critical value of about 4 nm and becomes metastable. A transformation into γ-Al_2O_3 then takes place. This phase transformation can reduce the thickness and the diffusion resistance of the oxide layer. For thin oxide layers observed at high heating rates, this phase change can also result in local discontinuities in the oxide coverage. The resulting γ-Al_2O_3 is stabilized due to its small crystallite size. This amorphous to γ-Al_2O_3 phase transformation can lead to a small mass increase because the density of the oxide layer increases. The growth of a continuous γ-Al_2O_3 layer and its partial transformation into θ-Al_2O_3

polymorph takes place in a third stage after the melting point of Al (660°C). Finally, the fourth stage corresponds to the high temperature oxidation above 1,100°C and leads to the formation of α-Al_2O_3. The transformation to thermodynamic stable α-alumina may proceed directly or via a number of intermediate phases such as δ- and θ-alumina.

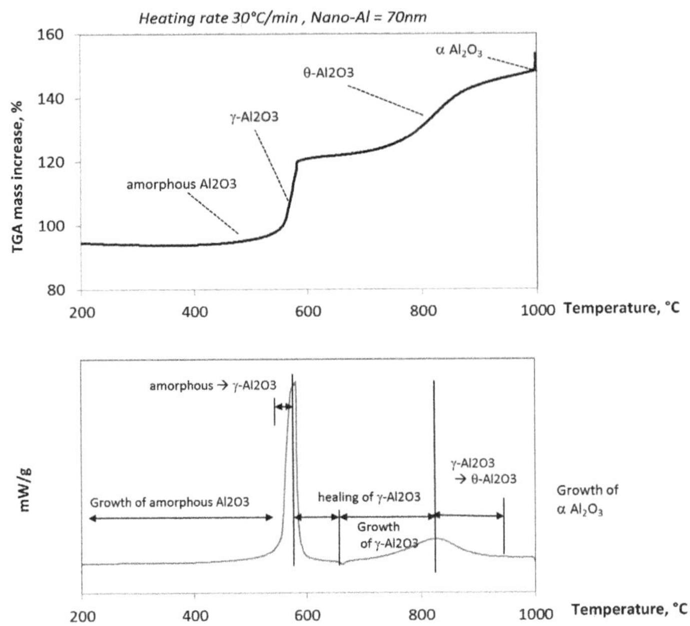

Figure 1.11. *An example of TGA curve showing the different aluminum oxidation stages (according to [TRU 06] description)*

The sequence of alumina phase transformations described above has made it possible to interpret the observed stepwise powder oxidation as pictured by experimental characterization such as a TGA experiment (see Figure 1.12). Five steps need to be analyzed to describe the oxidation quantitatively: (1) growth of the amorphous oxide, (2) the amorphous to γ-Al_2O_3 phase change, (3) growth of γ-Al_2O_3, (4) γ-Al_2O_3 to α-Al_2O_3 phase change and (5) growth of α-Al_2O_3. As an illustration, a TGA curve of the mass increase in an aluminum nanopowder heated in the thermal analyzer in argon and oxygen environment is given in Figure 1.12 indexed with the different direct oxidative growth and phase transformations. At each step, the oxide polymorphs are mentioned.

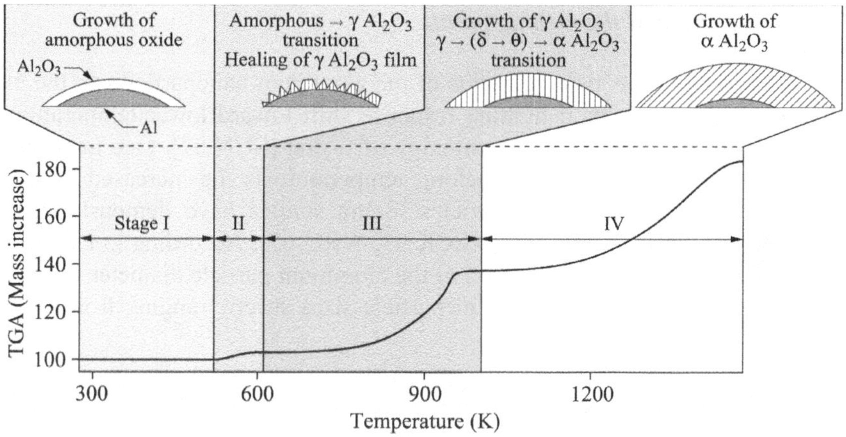

Figure 1.12. *Different stages of oxidation and the respective changes in the growing alumina scale are shown schematically [TRU 06]*

This model developed by Trunov *et al.* can predict that particles of different sizes ignite at different temperatures. The model consistently interprets a wide range of the previously published experimental data describing aluminum ignition.

1.5. Why incorporate Al nanoparticles into propellant and rocket technology?

It is now widely demonstrated that replacing the metal powders traditionally used in energetic mixture and rocket technology by Al nanopowders leads to a factor of 5–30 increase in the burning rate of propellants at a constant mass content of the components. This is primarily due to the following factors:

– increased reactivity of nanometer-size particles compared to micrometer-size particles;

– increased rate of energy release due to the lower melting point and much shorter combustion time of nanoparticles;

– favored mixing of components (aluminum and oxidizer), providing closer contact between them, which facilitates the diffusion of reactants to the surface and increases their reactivity.

1.5.1. *Reduction of the melting point*

Downscaling the average radius of the aluminum nanoparticles to the 5–50 nm range makes their melting response shift toward lower temperatures with noticeable decrease in the enthalpy of fusion [SUN 07]. One important factor in the reduction of Al melting temperature is the increased surface energy associated with nanoparticles. Some studies have demonstrated it theoretically and experimentally [ECK 93, WRO 67]. According to Eckert *et al.*, the melting point as a function of the aluminum particle diameter is given by the following empirical law for particle sizes strictly ranging from 12 to 43 nm.

$$T_m(in\ K) = 977.4 - \frac{1920}{d\ (in\ nm)} \quad [1.5]$$

For larger particles, $T_m = T_b$ with T_b bulk aluminum melting temperature = 660°C.

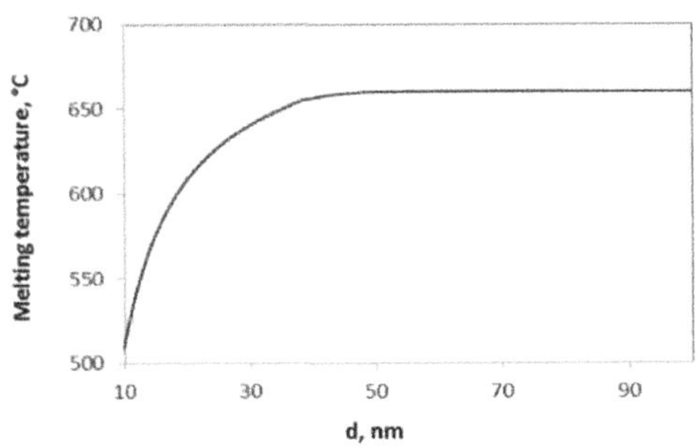

Figure 1.13. *Predicted melting temperature of aluminum nanopowders as a function of particles diameter (d)*

Decades ago [JON 00], Reiss and Wilson proposed a model describing the behavior of the melting point, T_m, as a function of the particle diameter d, and the oxide film thickness t_{oxide}, as:

$$T_m = T_b\left(1 - \frac{4\times\sigma_{SL}\times 1920}{H_b\rho_{Al}(d-2t_{oxide})}\right) \quad [1.6]$$

where T_b is the bulk aluminum melting temperature, H_b is the bulk aluminum enthalpy of fusion and σ_{SL} is the interfacial surface tension between the solid and liquid.

This model is often used and referenced in several recent papers dealing with aluminum nanopowder used in combustion systems [HUN 04, GRA 04]. The influence of the alumina passivation layer thickness has been studied by Sun and Simon [SUN 07]. The results indicated that the nature of the surface layer has no significant effect on the melting temperature.

1.5.2. *Increase in the reactivity*

It is observed that the reactivity of aluminum nanoparticles is considerably enhanced compared to micrometer-sized samples. Figure 1.14 illustrates the effect of aluminum size on the Al+O_2 reaction [SUN 06] where thermal analysis scans are given for 105 nm average diameter sample, 86 nm and 6 µm at a heating rate of 3°C/min under 25/75 O_2/Ar atmosphere. We clearly see that there is a large exotherm reaction peak present for the nanoparticles before reaching the Al melting peak at 660 °C. It is due to oxidation of the aluminum by O_2, whereas for the micrometer-sized particles, only a small exotherm is present prior to melting. The onset temperature for oxidation is reduced for the nanopowders. In addition, compared to a micrometer-sized sample, the melting peaks for the nanosized samples are considerably smaller, suggesting that more aluminum in the nanopowders reacted in the oxidation process prior to melting. Similar results were reported by Mench *et al.* in [MEN 98b].

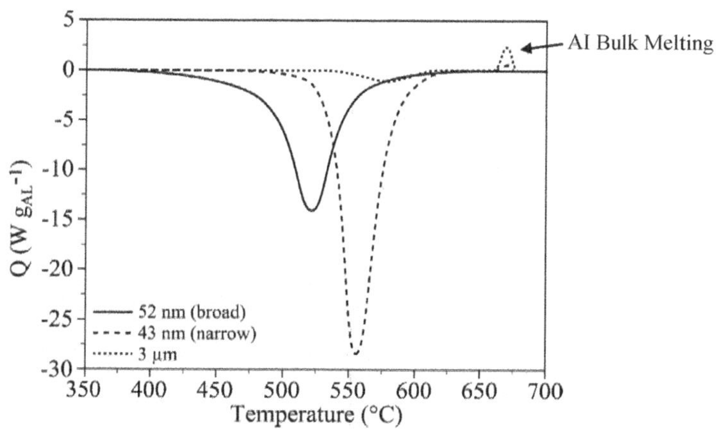

Figure 1.14. *DSC curves of aluminum powder as a function of the temperature in Al/O_2 environment showing the effect of the size and size distribution of the aluminum nanoparticles. The DSC scans were performed at 3°C/min under 25/75 O_2/Ar atmosphere [SUN 06] (Copyright 2006 Elsevier)*

The increased reactivity of nanoparticles is considered to be mainly related to their large specific surface area. Basically, when the particle size is reduced, the fraction of the atoms located at the surface increases. To illustrate this point, Figure 1.15 shows the surface to bulk atom ratio for a spherical aluminum crystal as a function of particle diameter. Surface atoms have free (unsaturated) bonds and a lower coordination, the bonds between surface and subsurface atoms are weaker than those between the atoms in the bulk. Atoms influence each other within approximately five interatomic distances. The thickness of this surface layer is about five atomic layers and is roughly equal to ≈ 1 nm.

Another effect typical of metal nanoparticles obtained by vapor-phase condensation technology is due to the presence of additional or excess of energy. This is due to the extremely fast formation of particles during condensation of metal vapor and the related formation of metastable and tense structures in the crystal lattice.

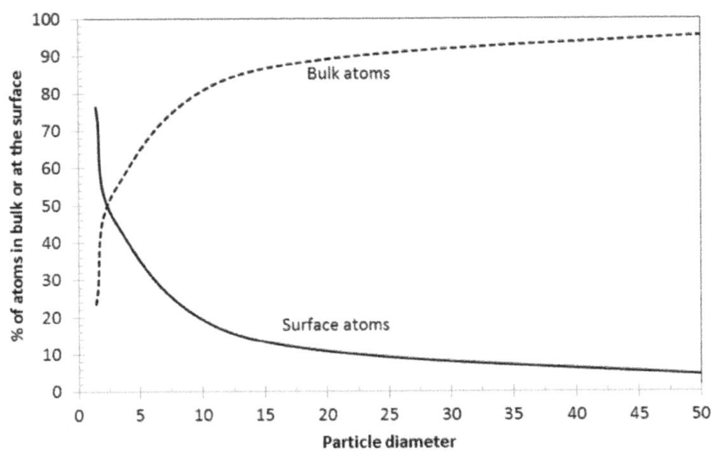

Figure 1.15. *Calculated percentage of atoms in aluminum particles of different diameters (diameter of an aluminum atom is equal to 0.286 nm)*

2

Applications: Al Nanoparticles in Gelled Propellants and Solid Fuels

One application of Al nanopowders concerns their incorporation into propellants, explosives and pyrotechnics to promote combustion, and, more specifically, to increase the burning rate. Propellant burning rate is in direct relation with the system pressure. *Pr*, the burning rate, is linked by the so-called Vielle's power law:

$$r = \alpha P^{\beta} \qquad [2.1]$$

where α and β are empirical and adjustable constants.

2.1. Gelled propellants

Much of the experimental and theoretical research on metalized gelled propellants began in the 1950s and 1960s with the study of slurry fuels considering various metals for aerospace applications [PAL 98]. In the 1970s, hydrocarbons, hydrazine derivatives and inhibited red fuming nitric acid emerged. In the last decade, most of the focus on metalized gelled propellants has concerned bipropellant systems [PAL 96]. Only very few of these studies have been dedicated to nanoscale particles (with diameter < 100 nm) [PAL 04]. Because of the high specific surface area of nanoscale gellants, gelled cryogenic propellants with nanoscale gellants require 25–50% less mass of gellant compared to traditional micrometer-size gellants. Recent investigations were conducted by various groups focusing on the use of nanosized aluminum as a gelling agent. Spray combustion of gelled rocket propellant (RP) fuel with Al nanopowder (ALEX®) was studied in a rocket motor with gaseous oxygen [MOR 01]. The mass fraction of ALEX® added to the RP was as high as 55%, although the mixture with 5% in weight had the best efficiency. A similar study [ELL 03], with 16% in weight of Al nanoparticles, showed good ignition and stability characteristics and efficiencies were doubled. Metalized gelled nitromethane, using ALEX® and 5 μm diameter particles, has also been studied in strand burner experiments [WEI 05]. Using 5% Aerosil (fumed silica) and ALEX® nanoparticle loadings of 5% and 10%, burning rates of approximately 4 and 5 mm/s were found at the highest pressure studied (~12–13 MPa), significantly higher than those of pure nitromethane. From theoretical calculations, it was determined that flame temperatures (confirmed by experiments) and specific impulses are also increased with the addition of Al nanopowder to gelled nitromethane.

Ivanov *et al.* [IVA 94, IVA 00] first investigated Al nanopowders in a mixture of water in the presence of polyacrylamide (3%) as a thickening agent. The particle diameter ranged from 30 to 100 nm. In their experiments, they mixed Al nanopowder with distilled water and added the thickening agent at equivalence ratios of 0.67 and 1.0. The mixture did not ignite in the absence of the polyacrylamide thickening agent. The mixture was filled into 10 mm tubes and ignited with an electrical coil inside a vessel maintained at constant pressure with argon as the atmospheric gas. At the maximum test pressure of 7 MPa, the maximum burning rate of the mixture was found to be approximately 1.5 cm/s. Shafirovich *et al.* [SHA 06] investigated the combustion behavior of 80 nm Al in water mixtures also with a polyacrylamide gelling agent. They found that 80 nm Al–H_2O mixtures yielded a combustion efficiency of ~50%. Risha *et al.* [RIS 07] investigated the combustion of Al nanopowder and liquid water without the use of any additional gelling agent. Steady-state burning rates were obtained at room temperature (25°C) using a windowed vessel for a pressure range of 0.1–4.2 MPa in an argon atmosphere with particles having a nominal diameter of 38 nm. The effects of particle diameter (50, 80 and 130 nm) and overall mixture equivalence ratio (0.5–1.25) on the burning rate were also studied at a pressure of 3.65 MPa. Chemical efficiencies in terms of specific impulse and burning rate were found to range from 27% to 99% depending upon particle size and sample preparation. Burning rates ($r = \alpha P^{\beta}$) increased significantly with decreasing particle size making it possible to reach rates as high as 8 cm/s for the 38 nm diameter particles under approximately 4 MPa. Pressure exponents, β, were determined to be equal to 0.47, 0.27 and 0.31 for the 38, 80 and 130 nm diameter particle mixtures, respectively.

Because of their high energy density, aluminum water mixtures have been considered as sources of hydrogen for power generation [SHA 06] and space propulsion [SIP 08, ING 04]. In the last decade, one study has considered Al nanoparticles with ice rocket motors [PER 07b].

DEFINITION (Gelled propellants).– *Gelled propellants are recognized as high-performance propellants in the sense that they are expected to share some of the advantageous properties of liquid and solid propellants taken separately. The gelling of liquid mono- and bipropellants reduces risk of leakages while maintaining its ability to be pumped and throttled, unlike solid propellants. Gelled propellants are generally less sensitive to impact, friction and electrostatic discharge than solid propellants. From a performance standpoint, gelled propellants offer high specific and density impulses, which are comparable to liquid systems, and performance may be further increased with more energetic materials such as metal particles.*

2.2. Solid propellants

The addition of aluminum nanoparticles in solid propellants has been proven in many studies to significantly enhance the burning rates of the propellant compared to the same propellant formulations with micrometer-size particles. Ivanov and Tepper [TEP 96] reported that by adding aluminum nanoparticles in a propellant formulation, the mixture burning rate could be enhanced by a factor of 5–10. Chiaverini *et al.* [CHI 97b] demonstrated an increase of 70% in burning rate of hydroxyl-terminated polybutadiene (HTPB)-based solid propellant doped with 20% of nanoparticles of Al. Armstrong *et al.* [ARM 03a] confirmed that the burning rate of the ammonium perchlorate (AP)-based propellants increases when conventional Al powder is replaced by Al nanopowder. The burning rate increases from 1 to more than 100 mm/s when the aluminum particle size decreases from 10 µm to 100 nm. Several experiments are performed at atmospheric and higher pressure conditions with the view to produce comparative combustion in a high-pressure closed vessel of a solid propellant containing 15% of either conventional micrometer-size Al or nanosized Al. This tendency has been confirmed for several types of propellants (HTPB/AP, glycidyl azide polymer (GAP)/AP and GAP/ammonium nitrate (AN)) and by different research teams [SIM 99, PIV 04, MEN 98a]. Mench *et al.* [MEN 98b] studied the effect of replacing a portion of the aluminum (50%) with Al nanopowder and observed an increase in the burning rates by a factor of 2, achieving rates near 1.8 cm/s. They showed that Al nanoparticles are a very effective burning rate enhancer for solid propellants and fuels.

Figure 2.1 shows the burning rate as a function of aluminum particle size published by Armstrong *et al.* and in a pressure ranges for the vertical lines.

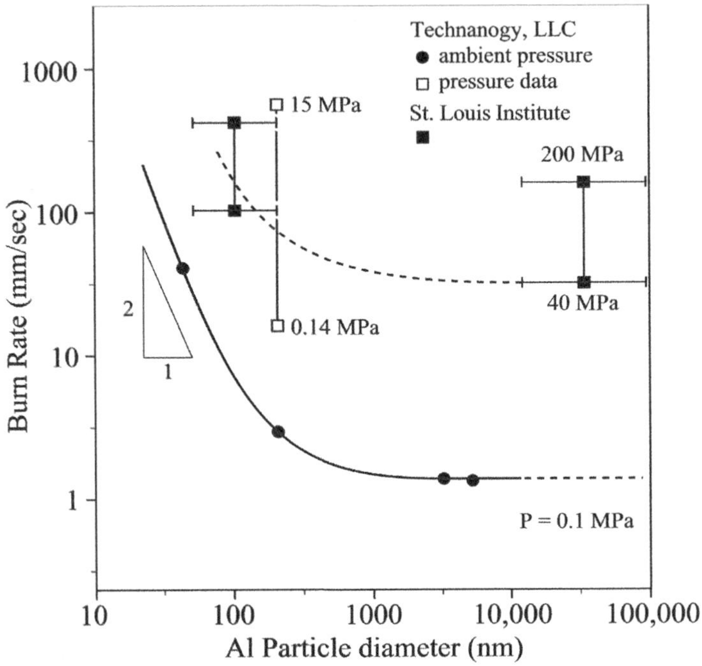

Figure 2.1. *Burning rate as a function of aluminum particle diameter at different pressure [ARM 03a] (Copyright 2003 American Chemical Society)*

While the use of nanoparticles can significantly increase burning rates and combustion efficiency, their high surface area can also introduce other complications particularly during processing. For example, propellants with only nanoparticles versus a mixture of micro- and nanoparticles can show unstable combustion because of the non-uniformities in the distribution of the nanoparticles during the fabrication process inducing local spots of high concentrations. The resulting propellants can be more brittle too, leading to increased sensitivity to cracking. To minimize these deleterious effects, micro-particles, including both oxidizer and fuels, are under development, including modifications to the particle surfaces, e.g. passivation layers to stabilize against unwanted ignition, to reduce long-term aging and to avoid surface contamination.

2.3. Solid fuel

As for solid propellant, many studies have proven that the addition of aluminum nanoparticles in solid fuel significantly enhances burning rates of the propellant compared to the same propellant formulations with micrometer-size particles only. Specifically, hybrid motor data showed that the addition of 13% of aluminum powders can increase the linear burning rate of solid HTPB-based fuel by 123% in comparison to the non-aluminized HTPB fuel and at a moderate gaseous oxidizer mass flow rate. Strand burner studies of two solid propellant formulations (one with 18% regular aluminum powder and the other with 9% aluminum replaced by ALEX® powder) showed that nanoparticles can increase the burning rate of solid propellants by 100%. The use of nanoparticles allows for energy release closer to the fuel surface, thus increasing the energy feedback rate to the surface and the regression rate of the fuel. Chiaverini *et al.* [CHI 97a, RIS 03] used a two-dimensional motor to study the effects of various weight percentages of ALEX®-type of aluminum nanoparticles (4%, 12% and 20%) in HTPB-based solid fuels. The addition of 20% ALEX® enhanced mass burning rates up to 70% in comparison to pure HTPB fuel formulations. Risha *et al.* [RIS 03] evaluated the addition of various Al nanoparticles in a long-grain center perforated hybrid rocket motor. ALEX® particles coated with Viton-A had the highest percentage increase in fuel mass burning rate with an enhancement of 120% compared to pure HTPB.

3

Applications of Al Nanoparticles: Nanothermites

The thermite reaction discovered in 1893 and patented in 1895 by the German chemist Dr. Hans Goldschmidt is a highly exothermic reaction that involves a metal (fuel or reducing agent) reacting with a metallic or a nonmetallic oxide (oxidizer) to form a stable oxide and the corresponding metal or nonmetal of the reactant oxide. This is a form of oxidation–reduction reaction that can be written as:

$$M + AO \rightarrow MO + A + \Delta H \qquad [3.1]$$

where M is a metal or an alloy, A is either a metal or a nonmetal, MO and AO are their corresponding oxides and ΔH is the enthalpy of reaction also called heat of reaction given in cal/g, J/g or J/cm^3. Several other parameters are associated with thermite reaction and commonly used: the ignition temperature also called onset temperature and ignition energy.

Ignition temperature corresponds to the temperature at which the exothermic reaction starts and is sustained without adding external energy.

Ignition energy corresponds to the thermal, mechanical or electrical energy required to bring the thermite mixture to its ignition point.

Reactivity corresponds to the decomposition rate, or reaction rate. The "traditional" thermite reaction is taken as the reaction of a stoichiometric mixture of aluminum and iron oxide (Fe_2O_3) reacting exothermically with the products being Al_2O_3 and Fe.

Figure 3.1. *Photographs of a thermite mixture made up of iron oxide and aluminum and its reaction (source: from Wikipedia: http://fr.wikipedia.org/wiki/Thermite)*

In addition to Al+Fe_2O_3, many thermite mixtures exist and have been studied. Fischer and Grubelich provide thermodynamic analysis of many exothermic thermites in [FIS 98]. Some of these reactions produce little or no gas. Others produce significant amounts of gaseous products. One thing to note is that these calculations assume that full density is achieved.

Table 3.1 gives the theoretical enthalpy of reaction and adiabatic temperatures under stoichiometric condition for a few selected thermite compositions used for energetic materials application.

Thermite composition	Stoichiometric mass ratio	Enthalpy of reaction ΔH (cal/g)		Adiabatic temperature T_{ad} (K)	
		COACH [COA 07]	Reference [FIS 98]	Without phase change	With phase change
4 Al + 3 MnO_2	1/2.147	1,146	1,159	4,829	2,918
2 Al + 3 CuO	1/4.422	987.8	974.1	5,718	2,843
2 Al + Fe_2O_3	1/2.959	952.0	945.4	4,382	3,135
2 Al + 3 NiO	1/4.454	855.1	822.3	3,968	3,187
4 Al + 3 SnO_2	1/4.189	686.8	686.8	5,019	2,876
4 Al + 3 SiO_2	1/1.670	515	513.3	2,010	1,889
Al + TiO_2	1/2.221	379.8	365.1	1,955	1,752

Table 3.1. *Stoichiometric mass ratio, enthalpy of reaction and adiabatic temperature for main thermites*

Thermite mixtures made up of nanosized components are termed nanothermite, MIC (for Metastable Intermolecular Composites or Metastable

Interstitial Composites), metal-based reactive nanomaterials or superthermites. Given the availability of Al nanopowders, aluminum is the most studied fuel in nanothermite systems with different oxidizers: Fe_2O_3, MoO_3, CuO, Bi_2O_3, MnO_2, WO_3, Ag_2O and I_2O_5. Fluoropolymer as polytetrafluoroethylene (PTFE) has also been tested as an oxidizer. The well-known brand name of PTFE-based formulas is Teflon by DuPont Co. [PAN 09a]: this last reaction consists of $4\ Al + 3(C_2F_4) \rightarrow 4\ AlF_3 + 6C$.

The study of nanothermites as energetic materials began at Los Alamos National Laboratory in the 1990s [AUM 95]. In this work, powder mixtures of Al and MoO_3 with average particle sizes of 20–50 nm were found to react more than 1,000 times faster than macroscale thermite material mainly due to reduced diffusion distances between individual reactant species. Interestingly, the enthalpy of a stochiometric mixture of Al and CuO can reach 4 kJ/g, which is greater than TNT (trinitrotoluene: $C_7H_5N_3O_6$). Since 1995, a considerable amount of research has continued in the area of nanothermites investigating the effects of particle size, oxidizer and fuel types, mixture ratio, packing density, pressure and mixture sensitivity. An introduction to many of these studies can be found in [ARM 03b, SON 07a, ROS 10] and [ROS 14]. In brief summary, the pressures and flame velocities are in between those of propellants and explosives: 10s and 1,000s m/s, depending on the confinement, while the pressures range between 1 and 10 MPa [MAR 11]. Extensive experimentations also demonstrated that the maximum gas pressure and reactivity can be tuned just by changing the particle size, particle concentration, the uniformity of mixing, the choice of oxidizer and, of course, the stochiometry and mixture compaction [KWO 13, SUL 12c, WEI 09].

In the following, the first subsection presents the methods of preparation for nanothermites and gives some examples of realization. Then, a comprehensive discussion on the main influencing parameters in their ignition and combustion properties is proposed. A final subsection details the different experimental means developed and conventionally used to characterize nanothermite reaction behaviors and performances.

3.1. Method of preparation

In nanothermites, the oxidizer (mostly metallic oxide) and the reducing agent (also called fuel) are intimately mixed in the nanometer scale. Different methods have been proposed and developed to date. The most common preparation technique is the physical mixing of nanoscale ingredients, i.e. nanoparticles of oxide and aluminum which will be presented first. Other techniques will be described such as rapid expansion of a supercritical dispersion (RESD) or molecular self-assembly methods.

3.1.1. *Ultrasonic nanopowder mixing*

Ultrasonic mixing in a bath of solvent (hexane, isopropyl alcohol or another liquid carrier) [PAN 05, PLA 05, SON 07b, PUS 07] which is subsequently evaporated is the traditional and simplest method to prepare a nanothermite powder. The Al and oxidizer nanopowders are weighted and suspended in the solvent in a beaker. The beaker is then placed in an ultrasonic bath of water. After a certain time of ultrasonic agitation, the colloidal suspension is dried at low temperature under low pressure (e.g. 70°C in the case of hexane) and the mixture is collected for characterization or use. The sonication lasts for a total duration of a few minutes. For example, 1 s delay is used after every 2 s of sonication to avoid overheating of the solution that could lead to undesired ignition of powder. For liquids that can oxidize aluminum, such as isopropanol, the exposure time to the solvent is reduced.

The assessment of the quality of mechanical mixing of nanopowders, related to the sample preparation method, is of crucial importance while being difficult to be evaluated quantitatively. The different imaging and thermal analysis techniques, presented in preceding section, are used for characterizing the mixtures mixing uniformity: scanning electron nicroscopy (SEM), high-resolution transmission electron microscope (HR-TEM) and, to a lesser extent, electron energy loss spectroscopy (EELS).

For example, SEM images presented in Figure 3.1 show the distribution, shape and different sizes of the particles in four different nanothermites obtained by powder mixing [GLA 14] using ultrasonication in hexane. The characteristics of oxidizer powders used to make these nanothermites are given in Table 3.2.

Materials	Manufacturer	Measured size	Shape
Al	Novacentrix	80–150 nm	Spherical
Bi_2O_3	SigmaAldrich	100–500 nm	Ovoid
MoO_3	SigmaAldrich	90–6,000 nm	Random shape
CuO	SigmaAldrich	240 nm +/– 50 nm	Random shape
PTFE	SigmaAldrich	800–1,500 µm	Spherical

Table 3.2. *Summary of powders used in the nanothermite mixture*

SEM of the Al/CuO nanothermite showed that the CuO is really non-spherical. The average dimension is 240 nm with a dispersion of 50 nm. For the other particles, there are different ranges of size. The bismuth trioxide particles are ovoid with a dimensional average of 600 nm × 400 nm for the biggest and 110 nm × 90 nm for the smallest. The molybdenum trioxide particles shape is randomly distributed with a dimensional average of 2 μm × 6 μm for the biggest and 90 nm × 90 nm for the smallest size. These SEM images clearly illustrate the variety of morphologies for nanothermite materials obtained by powder mixing since initial powders are of different shapes. These observations underline the issue of mixing quality.

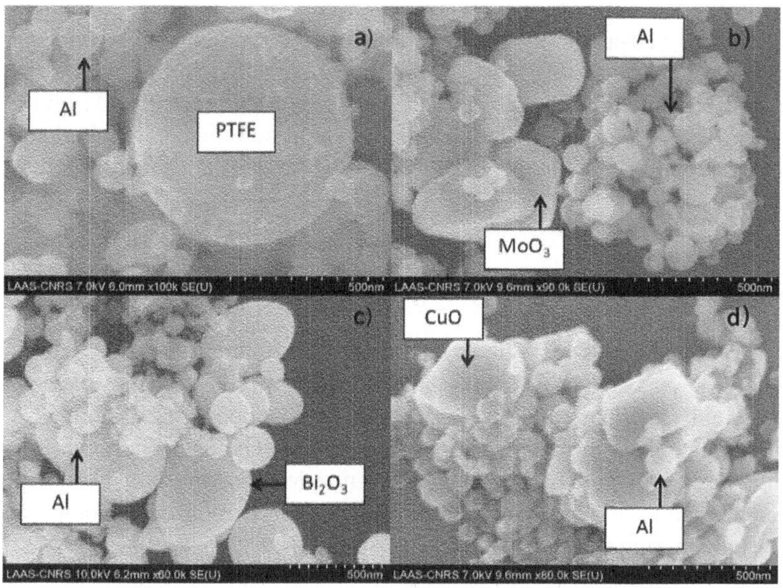

Figure 3.2. *Scanning electronic microscopy images of the different nanothermite prepared by a) powder mixing aluminum/PTFE (Al/PTFE); b) aluminum/molybdenum trioxide (Al/MoO_3); c) aluminum/bismuth trioxide (Al/Bi_2O_3) and d) aluminum/copper oxide (Al/CuO) (Copyright 2015 Elsevier)*

In summary, mixing nanopowders using the ultrasonication method is simple and widely used in the laboratory but faces several drawbacks leading to scattering of reaction performance and decrease in operation reliability, as follows:

– it is difficult to scale up. The processing of larger sample batches inevitably leads to a lower quality of mixing;

– the ultrasonication used to disperse agglomerates results in increased mixing intimacy but segregation of particles during drying forms undispersed agglomerates leading to inhomogeneities. Moreover, aluminum nanoparticles are passivated with a thin oxide shell that degrades interfacial contact between oxide and aluminum but also permits to improve mixture stability at low temperature.

Some more sophisticated processes have been investigated to improve the mixing quality. Among them, Marioth *et al.* [MAR 06] mix Al and Fe_2O_3 nanopowders using supercritical dispersion. More recently, self-assembly techniques have been vaunted as a cutting-edge solution for precise organization. Al/Fe_2O_3 nanothermites were, for instance, obtained by electrostatic self-assembly based on oppositely charging of Al and Fe_2O_3 nanoparticles in aerosol [KIM 04] or by functionalization of Al and Fe_2O_3 nanoparticles with oppositely charged ligands. More recently, Al and CuO nanoparticles have been arranged in inorganic solution or manipulated by using a polymers and bioagents [SEV 12, KIM 04], as described in following sections.

3.1.2. *Rapid expansion of a supercritical dispersion*

The technique is as follows: as supercritical fluids have no interfacial tension, they can be used for the homogeneous mixing of nanoparticles that are normally agglomerated in a dry state. The nanoparticles are mixed in an agitated autoclave in a supercritical atmosphere. The dispersion can then be either rapidly or slowly expanded by a nozzle, breaking the agglomerates. This technique allows to achieve a better degree of mixing compared to ultrasonic mixing.

Figure 3.3. *Photograph of the rapid expansion of a super critical dispersion (RESD) at ICT Franhaufer*

The main concern in developing and scaling up RESD for mixing reactive nanopowders appears to be the safety of operation. In particular, the reactions between the supercritical fluid and Al nanoparticles need to be prevented.

3.1.3. *Molecular self-assembly of nanoparticles*

Molecular self-assembly has been investigated recently by a few teams to control the fuel/oxidizer interfacial contact area and improve the homogeneity of the nanothermite mixture. Self-assembly means that the Al nanoparticles can arrange by themselves, or directed by external forces, in a controlled manner around the oxidizers or vice versa. It can be done in inorganic, aqueous or organic solutions. Kim *et al.* [KIM 04] proposed a method in which the Al and Fe_2O_3 self-assembly is controlled by the electrostatic forces which exist between charged aerosol particles. The authors compared the reactivity of resulting assembled Al/Fe_2O_3 with a randomly assembled Al/Fe_2O_3. Characterizations give a total enthalpy of reaction of 0.7 kJ/g over a relatively wide temperature range for Brownian assembled particles, while a total enthalpy of 1.8 kJ/g over a very narrow temperature range was observed for the electrostatically assembled case.

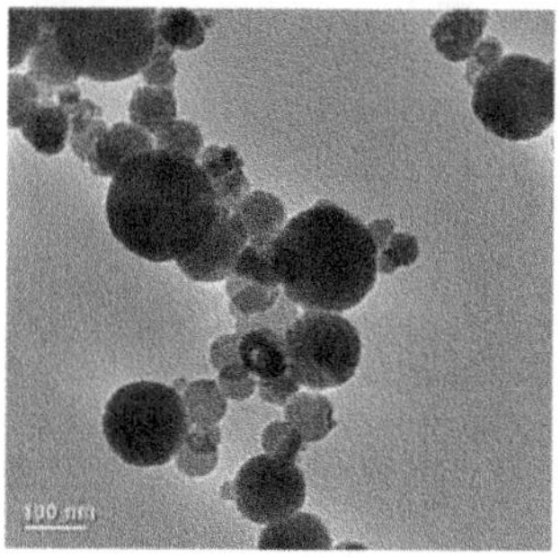

Figure 3.4. *Transmission electron microscopy (TEM) images of self-assembled Al/Fe_2O_3 nanoparticles based on electrostatic forces [KIM 04] (Copyright 2004 Wiley)*

Malchi *et al.* [MAL 09] proposed an assembly method based on electrostatic forces between two ligands (see principle in Figure 3.5). Two ligands (with a positive and negative ω-functionalization) are attached to the surface of Al and CuO nanoparticles to create a charged self-assembled monolayer (SAM). The ligands used to functionalize the surfaces of the aluminum particles are a ω-trimethylammonium (TMA) functionalized carboxylic acid, HOOC $(CH_2)_{10}$ NMe_3+Cl^-. The CuO nanoparticle surfaces are treated with the ω-carboxylic-acid-functionalized thiol, or mercaptoundecanoic acid (MUA), $HS(CH_2)_{10}COOH$. Experiments with the self-assembled material demonstrated the ability to be ignited and to sustain a combustion in microchannels, which does not occur with material having similar amounts of organics mixed via the traditional sonication method.

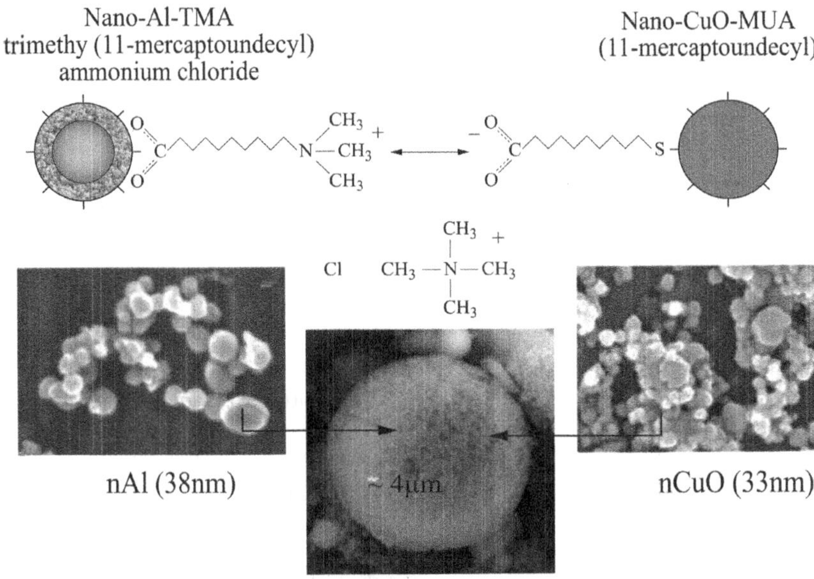

Figure 3.5. *Schematics of the assembly principle and SEM images of self-assembled Al/CuO nanothermites based on electrostatic forces which exist between two ligands [MAL 09] (Copyright 2009 American Chemical Society)*

Severac *et al.* [SEV 12] reported a DNA-based assembly protocol of Al and CuO nanoparticles. Two distinct grafting strategies have been followed to bind oligonucleotides on Al and CuO nanoparticles, respectively: thiol-modified oligonucleotides are attached to CuO nanoparticles and a biotin-(neutr)avidin system is used to coat Al nanoparticles. In the first case, the strong affinity of thiol groups for CuO allows us to directly graft thiol-modified oligonucleotides to CuO nanoparticles. In the second case, neutravidin is adsorbed first on the thin alumina shell covering the aluminum nanoparticles, and biotin-modified oligonucleotides are then grafted on these protein-modified nanoparticles. The assembly process is illustrated in Figure 3.6. The authors have characterized the resulting Al/CuO nanothermites (see SEM photograph in Figure 3.7) by differential scanning calorimetry (DSC) showing a total enthalpy of reaction (ΔH) of 1.8 kJ/g which is among the best ever achieved. They have also demonstrated the possibility to tune the onset temperature by changing the size of aluminum particles while keeping the same assembly protocol.

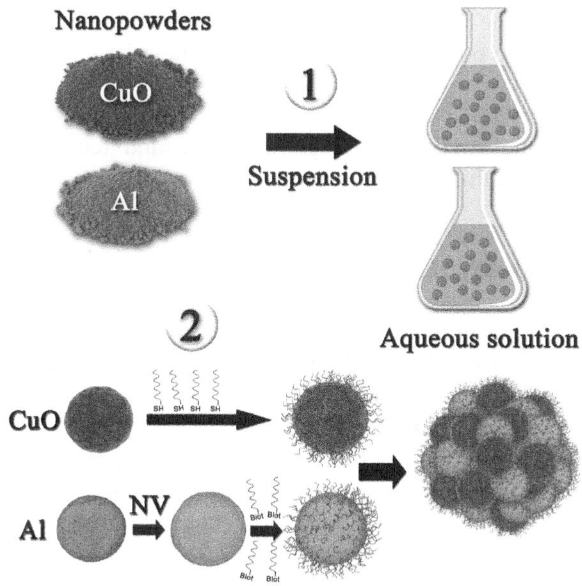

Figure 3.6. *Schematics of the different steps for the DNA-directed assembly of Al/CuO nanothermites. Al and CuO nanopowders are first suspended and stabilized in aqueous solution, and then functionalized with single DNA strands, and eventually assembled through hybridization of complementary DNA strands [SEV 12]*

Figure 3.7. *Scanning electron microscopy image of DNA-assembled Al/CuO aggregates based on forces which exist between two DNA strands [SEV 12]*

3.2. Key parameters

Whatever the method chosen to mix aluminum fuel and oxidizer (ultrasonic physical mixing or self-assembling method), a few parameters are important when preparing nanothermite before any characterization. In addition to the mixing uniformity, the mixture compaction, particle size, the stochiometry and the aluminum passivation layer must be well defined and controlled. All these points will be discussed in the following sections.

Other parameters also greatly impact the nanothermites ignition and combustion as the environmental gas pressure and temperature, but this won't be detailed in this book.

3.2.1. *The bulk density, theoretical density and compaction*

After the mixing process step, the powder mixture is often pressed into solid pellets using a standard hydraulic press. The die pressure level permits to create a range of bulk sample densities between 50% to 90% of the theoretical maximum density (TMD). The TMD is calculated as a weighted average of the pure solid densities of the different reactants of the mixture. The bulk density is given in g/cm^3. To obtain low bulk sample densities, the nanothermite mixture is not pressed. The compaction pressure is typically between 100 and 500 MPa to set the TMD percentage (% TMD) from 50% to 100% (fully dense pellet) [STA 11c].

Figure 3.8. *Schematics of the method to compact the nanothermite mixture and increase the TMD percentage parameter*

Nanothermite reaction rates are strongly affected by bulk density of the thermite. The lower the percentage of TMD is (i.e. the lower the compaction), the greater the burning rate will be. As the material pellet density is increased, the air volume within the interstitial void spaces is decreased. Eliminating the presence of air in the overall pellet will effectively increase the pellets thermal diffusivity. If a dominant mechanism for flame propagation is energy transport by diffusion, then higher thermal properties will enhance the burning rate as observed in Figure 3.9.

As seen in Figure 3.9, this trend is the opposite for micrometer-sized thermite mixtures. At low bulk densities for micron-sized mixtures, the reaction rates are low due to the larger length scales and void spaces, and heat losses represent a greater percent of the energy release.

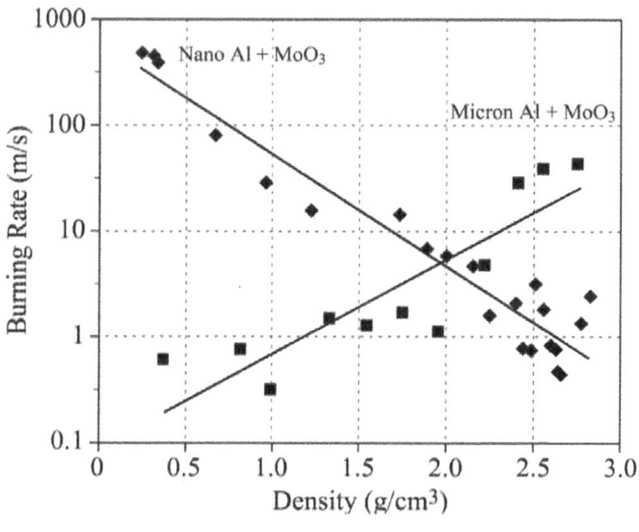

Figure 3.9. *Measured burning rate as a function of bulk density [PAN 05]*

In [STA 11b], Stamatis *et al.* measured that the ignition delay using a 50 W laser of Al/MoO_3 nanothermite powder increases with the increase in TMD percentage. The nanothermite powder was prepared by ultrasonic mixing in hexane of aluminum nanopowder having a nominal size of 80 nm with MoO_3 particles.

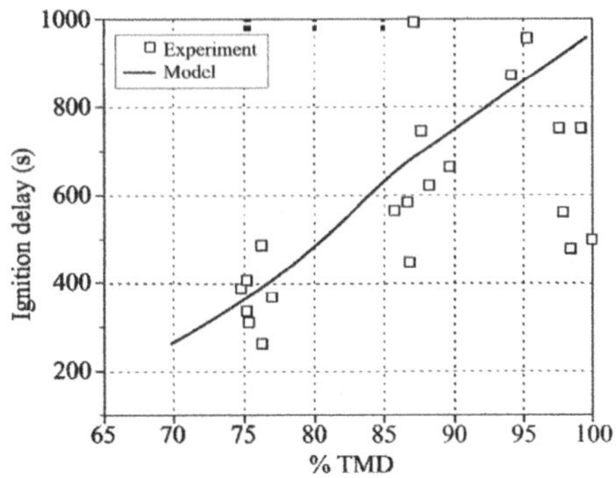

Figure 3.10. *Measured ignition delay as a function of Al/MoO$_3$ (8Al.MoO$_3$) pellet density using a laser power of 50 mW [STA 11a]*

Combustion mechanisms vary with compaction: at approximately 50% TMD threshold, the dominant thermal transfer mechanism transitions from convective to conductive burning due to the reduction of air impregnated in the nanoparticles.

3.2.2. *The stochiometry*

The stochiometry is the calculation of relative quantities of reactants in the chemical reactions leading to a complete reaction. In the literature about nanothermite, we conventionally talk about the mass equivalence ratio Ø. Considering the corresponding reaction: x Al + y Oxide, Ø is expressed as follows :

$$\varnothing = \frac{\left(M_{Al} / M_{Oxide} \right)_{SAMPLE}}{\left(M_{Al} / M_{Oxide} \right)_{ST}} \qquad [3.2]$$

where:

$$\left(Al / oxide \right)_{ST} = \frac{xM_{Al}}{yM_{Oxide}} \qquad [3.3]$$

In this equation, M_{Al} and M_{Oxide} are the molecular weight of aluminum and oxide, respectively. *SAMPLE* indicates the studied nanothermite ratio and *ST* indicates the stochiometric ratio. The stochiometric ratio depends on the thermite couple; some are reported in Table 3.3. In reality, it is important to remember that each Al nanoparticle is passivated with a thin alumina coating and a term of purity (*P*) has to be considered to obtain the fuel to oxidizer mass ratio in the mixture.

The mass ratios of fuel and oxidizer are therefore given by:

$$X = \frac{\phi.\left(^{Al}/_{Oxide}\right)_{ST}}{P+\phi.\left(^{Al}/_{Oxide}\right)_{ST}} \quad [3.4]$$

Materials	Stochiometric mass ratio
$4Al+3MnO_2$	1/2.15
$2Al+3CuO$	1/4.42
$2Al+Fe_2O_3$	1/2.96
$2Al+3NiO$	1/4.45
$4Al+3SiO_2$	1/1.67
$Al+TiO_2$	1/2.22
$Al+MoO_3$	1/2.67
$Al+WO_3$	1/4.30
$Al+Sb_2O_3$	1/5.40

Table 3.3. *Stochiometric mass ratio for different thermite couples*

The stochiometry of the mixture can be adjusted readily to maximize the energy density [SON 07b, SAN 07, BOC 05]. It should be noted that in many studies, reactions are performed in an ambient air environment such that oxygen from the surroundings may contribute to the reaction. In reality, the contribution from Al reacting with ambient air may shift this optimum point from fuel rich (excess of Al) toward stochiometric. However, the precise contribution of ambient air in the reaction is yet unknown. As a result, optimum thermite compositions are often determined to be fuel-rich, i.e. $\phi>1$.

Dutro *et al.* [DUT 09] investigated the effect of stochiometry on the combustion behavior of the nanoscale aluminum molybdenum trioxide thermite (Al/MoO_3) by characterizing the propagation velocity and pressure output of the reaction. The particles are spherical with a diameter of 80 nm. The MoO_3 used have a sheet-like morphology with typical dimensions of approximately 30 nm × 200 nm. The mixture ratios of the nanothermites varied more than 5% of nanoparticles of Al (95% MoO_3) and 90% of aluminum (10% MoO_3). Results revealed three separate combustion regimes:

a steady high-speed flame propagation (~100–~1,000 m/s) from approximately 10% to 65% content of Al, an oscillating and accelerating wave near 70% content of aluminum and a steady slow-speed flame propagation (~0.1–1 m/s) from approximately 75% to 85% content of aluminum. Propagation failed for the two extremes: fuel lean with mixture ratio <10% of aluminum and fuel rich >85% aluminum. The authors suggest that the propagation mode is a supersonic convective for near stochiometric mixtures and a conductive deflagration for extremely fuel-rich mixtures. This illustrates the great influence of the stochiometric ratio and the need to consider this parameter when preparing thermite mixture.

3.2.3. *The size of Al and oxidizer particles*

Both the size of the fuel and oxidizer particles are important. The effect of decreasing particle size is generally to increase the burning rate and lower the ignition temperature and energy. Eventually, for very small aluminum particles (i.e. below 50 nm in diameter), the burning rates become independent of particle size because the Al_2O_3 content percentage increases to become a larger portion of the total material. Weismiller *et al.* [WEI 11b] investigated the effect of Al and oxidizer particle sizes of Al/CuO and Al/MoO_3 thermites prepared by ultrasonic mixing in hexane. Particle size varied between the nanometer and micrometer scale. The equivalence ratios of the mixtures were those found to optimize the burning rate. Table 3.4 below summarizes the results.

	Linear burning rate (m/s)	Pressurization rate (MPa/μs)
Nano Al/nano CuO	980	0.67
Micron Al/nano CuO	660	0.28
Nano Al/micron CuO	200	1.82
Micron Al/micron CuO	180	0.11
Nano Al/nano MoO_3	680	0.68
Micron Al/nano MoO_3	360	0.20
Nano Al/micron MoO_3	150	0.44
Micron Al/micron MoO_3	47	0.17

Table 3.4. *Summary of experimental burning rate and pressurization rate of Al/CuO (Ø = 1) and Al/MoO$_3$ (fuel rich, Ø = 1.4) thermites [WEI 11b]*

It was found that reducing the size of the oxidizer particle has a greater effect on the burning rate than reducing the size of the Al particles. This is mainly due to the fact that decreasing the scale of aluminum particle induces an increase in the amount of Al_2O_3 in the mixture [MAL 08]. For aluminum nanoparticles of 50 nm in diameter, the amount of native Al_2O_3 reaches ~70%. This study concludes that there is a trade-off between reducing particle diameter and increasing Al_2O_3 content which has been confirmed by Pantoya *et al.* [PAN 05], suggesting that for Al particle sizes less than 50 nm in diameter, the attributes of a smaller Al particle size are overpowered by the inhibiting characteristics of an excessive Al_2O_3 content.

In another study, Granier *et al.* [GRA 04] compared the ignition time and burning rate of slightly fuel-rich Al/MoO_3 composites prepared by ultrasonic mixing (Ø = 1.2). Two mixtures were prepared: one from small particles (diameters of 108, 39 and 30 nm) and one with large particles (20 μm). A 50 W CO_2 laser was used to ignite the nanothermite mixtures prepared at 38% TMD. In this study, ignition delay corresponds to the time delay between the laser onset pulse and the first detection of the optical emission associated with the nanothermite reaction. As illustrated in Figure 3.11, decreasing Al particle diameter from micrometer to nanometer scale in an Al/MoO_3 mixture decreases the ignition time by two orders of magnitude. The increased sensitivity to ignition in nanometer Al particle mixture may be attributed to the Al reduction in melting temperature associated with nanometer particles (see section 1.5.1 and Figure 1.13).

The millisecond ignition delays could have been described theoretically by modeling the heat transfer processes of the pellet heated by the laser; however, such a description requires the value of the bulk thermal conductivity of the pellet, which is very difficult to either predict or measure. This difficulty is common for all nanothermite mixtures.

As illustrated in Figure 3.12, burning rates increase as the particle size increases with particle size except in the cases where the Al_2O_3 content exceeded 50% (for Al diameter <50 nm). The Al_2O_3 may decrease the burning rate and overpower the theoretical effect that reducing particle size will increase burning rate.

The absence of a clear effect of the size of aluminum nanoparticles on the ignition delay most likely results from the differences in the mixing quality of the thermite mixtures prepared with different nanopowders. It is, for example, reasonable to expect that the finer nanoparticles (smallest diameter) were mixed less uniformly with MoO_3 than coarser ones so that the effect of reduced particle size was effectively negated.

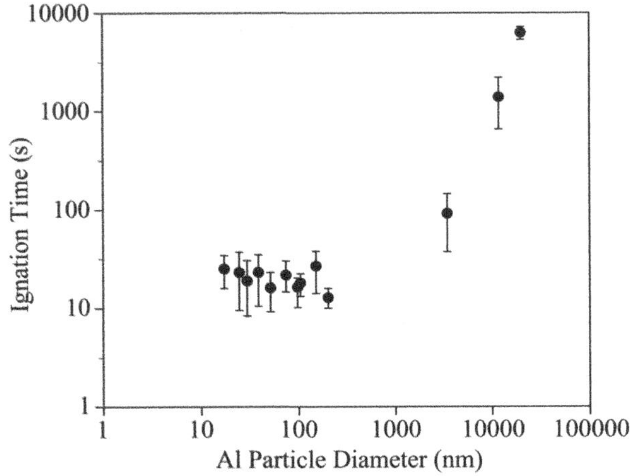

Figure 3.11. *Measured ignition time as a function of Al particle diameter plotted on a log/log scale (source: http://www.elsevier.com/wps/find/journaldescription.cws_home/505736/description#description) (Copyright 2004 Elsevier)*

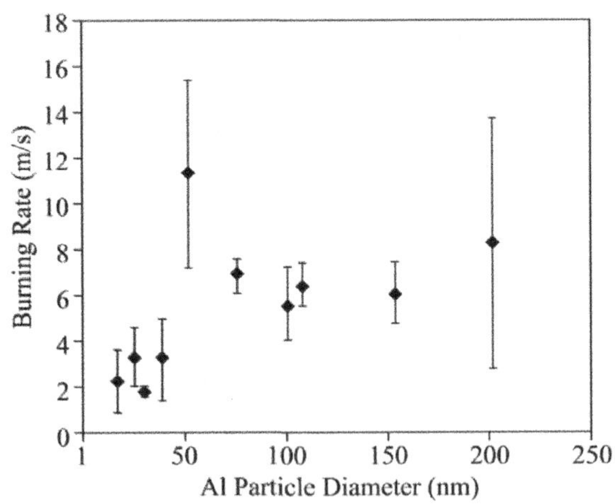

Figure 3.12. *Measured burning rate as a function of Al particle diameter plotted on a log/log scale (source: http://www.elsevier.com/wps/find/journaldescription.cws_home/505736/description#description) (Copyright 2004 Elsevier)*

3.2.4. *The passivation layer*

Jones *et al.* [JON 03] studied the effect of the thickness of the passivation layer on the reactivity of Al nanopowders in air using thermal analysis techniques. Three types of Al particles were considered: some Al nanoparticles with size in the range of 20 ± 50 nm (covered with Al_2O_3 passivation layer) and some from the same manufacturer coated with ~18 wt (weight)%[1] fluoropolymer. These nanoparticles are compared with ALEX®-type ones. Their results suggest that the thickness of the passivation layer does not significantly influence the reactivity of the material; but other factors play an important role such as particle size distribution, degree of agglomeration as well as the nature of the passivation layer. Although the reactivity of the particle may not be strongly influenced by the thickness of the oxide shell, the increased Al_2O_3 content of nanopowders may significantly influence the microstructure and macroscopic properties of the reaction.

The Al nanoparticles with sizes in the range of 20 ± 50 nm and covered with a Al_2O_3 passivation layer are less reactive in air than Al ALEX® nanoparticle with sizes of 180 nm, while it has a smaller particle size. Its lower reactivity in air may be due to a relatively thicker oxide layer resulting from aging of the sample. The coating with fluoropolymer significantly reduces its reactivity with water, which can consequently reduce the effect of aging. Chowdhury *et al.* [CHO 10] have experimentally studied the ignition of Al/CuO nanothermites with different aluminum oxide shell thicknesses. The ignition delay increases with an increase in shell thickness. The aluminum oxide shell is thickened from 2 to 4 nm by thermal treatment. The experiments were conducted under high heating rate. ALEX® particles with nominal size of 50 nm are mixed with CuO.

In [KAP 12], the authors studied the effects of aluminum surface functionalization on the burning rate for three different thermites mixtures consisting of aluminum particles, with and without surface functionalization, mixed with MoO_3. More precisely, two types of Al particles have SAMs of perfluorotetradecanoic (PFTD) and perfluorosebacic (PFS) acids around the alumina shell, respectively; the other one does not. Burning rates for Al with PFTD acid combined with MoO_3 are 86% higher than Al/MoO_3 (497 against 267 m/s), whereas those for Al with PFS acid combined with MoO_3 are

1 Weight%: Weight percent stands for the mass fraction with a denominator of 100. This is the percentage by weight.

almost half of Al/MoO_3 (138 against 267 m/s). The Al-PFTD structure is more sterically hindered and shows lower bond dissociation energy. This chemistry promotes increased burning rates. This study clearly shows that aluminum surface functionalization can be used as an approach to tune the reactivity of Al particles.

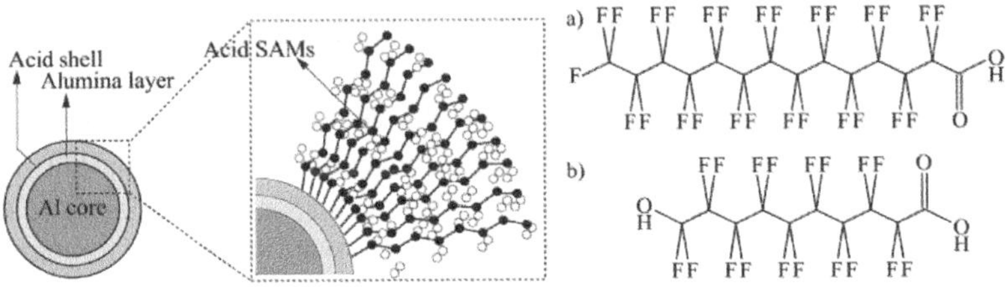

Figure 3.13. *Schematics of the chemical structure of PFTD and PFS and one particle surrounded with acid shell [KAP 12] (Copyright 2012 American Chemical Society)*

3.3. Pressure generation tests

There have been a few studies revealing that nanothermite can generate pressure impulses (see Table 3.4, pressurization rate column). Typically, the pressurization rate is determined by measuring the pressure generated during the reaction front propagation as a function of time. Measurements can be done in different confinement and environments. Figure 3.14 shows a typical experimental setup for pressurization measurements of nanothermite powder. Typically, a sample of a few milligrams of nanothermite is loaded in a cylindrical metal vessel, which is called the combustion chamber. To increase the packing density, the powder can be pressed by applying pressure. The nanothermite powder is then ignited directly with a metallic wire in contact at the top. Piezoelectric-based pressure sensors are used to measure the pressure–time characteristic.

In [GLA 14], Glavier *et al.* have characterized overpressure generation of various kinds of Al-based nanothermites (Al/CuO, Al/Bi_2O_3, Al/Fe_2O_3 and Al/PTFE) prepared at 30% TMD. Al nanoparticles are the same for all nanothermite systems: 80 nm ALEX® Al nanopowder covered with a 2.5 nm-thick amorphous aluminum oxide shell.

As illustrated in Figure 3.15, all nanothermites produce pressure. The maximum pressurization rate for Al/Bi_2O_3 mixture is ~5,762 kPa/μs, comparatively higher than the others: 172, 35 and 33 kPa/μs for Al/CuO, Al/MoO_3 and Al/PTFE, respectively. The gas is generated by the decomposition and vaporization of the oxide particles. In the case of Al/CuO, this corresponds to the conversion of CuO into Cu_2O with the release of oxygen first and then the vaporization of Al and Cu at higher temperature. In the case of Al/MoO_3 and Bi_2O_3, the gas generated is thought to be gaseous MoO_3 and Bi_2O_3 that can decompose into Bi_2 and O_2 at high temperature, respectively.

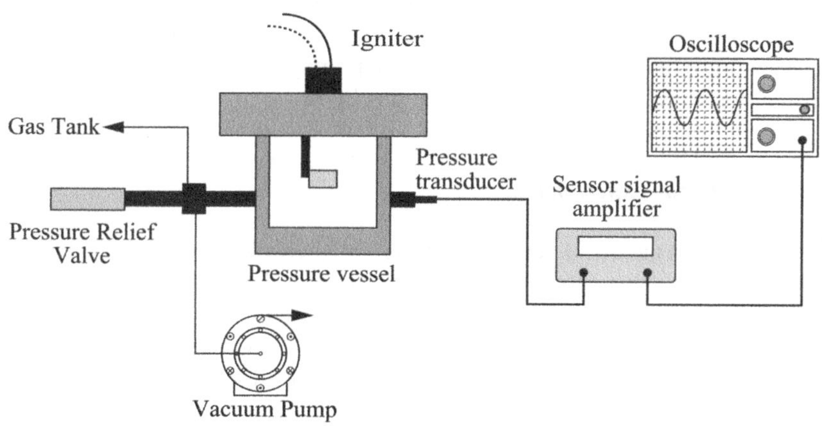

Figure 3.14. *Schematics of the experimental setup for pressure–time measurements*

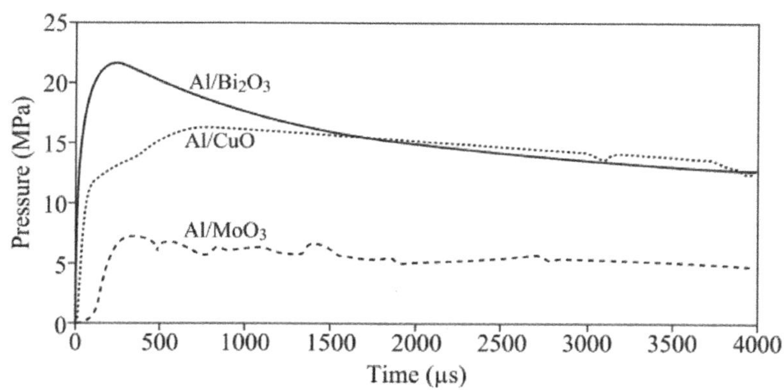

Figure 3.15. *Measured pressure as a function of the time for different nanothermite (Ø = 1) reaction at 30% TMD*

3.4. Combustion tests

Thermite reactions are highly luminescent and the significant light intensity makes imaging with a standard high-speed camera possible. There is a large number of "qualitative" or "semi-quantitative" measurements of reactivity using different techniques and in different environments: confined or unconfined as described in the following sections. Combustion tests permit to measure flame propagation velocity, also called burning rate, self-propagating reaction rate, flame speed or propagation rates. These different terminologies can be found throughout this book.

3.4.1. *Open tray experiments*

A mass (typically a few milligrams) of thermite is placed in an open sample holder, most often an elongated channel. The powder is then ignited directly with a metallic wire in contact at an extremity or by a laser, and the propagation of the flame is typically visualized using high-speed video. Photodiodes can be added. A schematic of an open tray combustion test is given in Figure 3.16

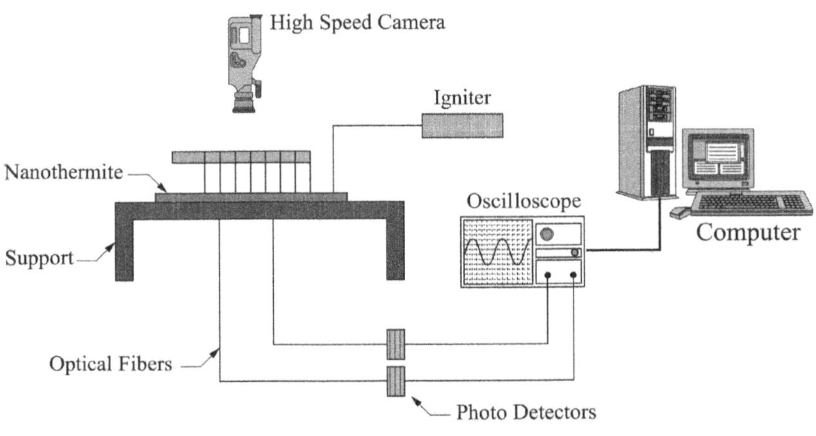

Figure 3.16. *Schematic of an open tray burning rate measurement setup*

A photograph of such an experiment, presented in [GLA 14], is shown in Figure 3.17. Figure 3.18 provides video images of the flame-front of slightly fuel-rich Al/CuO nanothermite (∅=1.1) prepared by ultrasonic mixing and not being compacted.

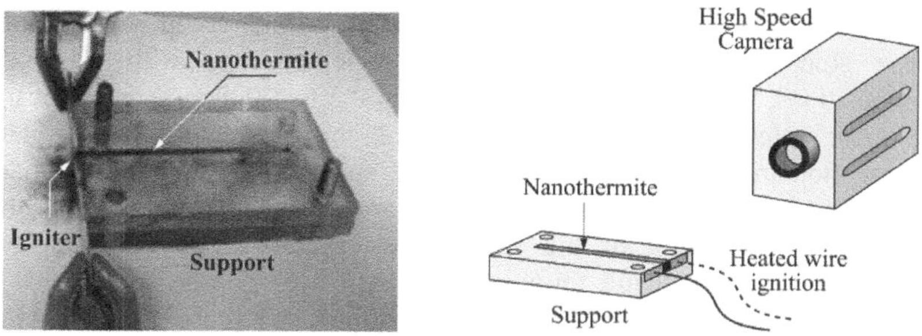

Figure 3.17. *Photograph of open channel apparatus used for measuring flame velocity (burning rate)*

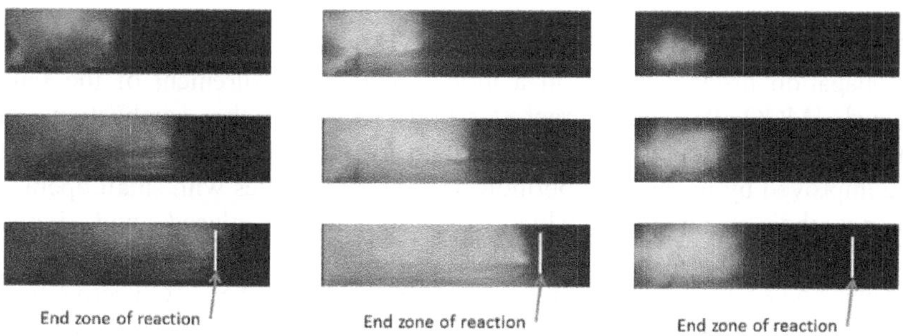

Figure 3.18. *Sequence of high-speed video frames of the combustion of different nanothermites prepared by powder mixing in hexane and not compacted. Images were recorded at 20 μs intervals and two images are separated by 40 μs between them [GLA 14]. Left: Al/Bi_2O_3, middle: Al/CuO, right: Al/MoO_3 (Copyright 2015 Elsevier)*

Open tray experiments are the most instrumented to obtain more quantitative and meaningful combustion characteristics. As already mentioned, and as for ignition, the flame speed or burning rate is strongly affected by the particles size distributions, the type of nanothermites [PAN 05, MOO 04, PLA 05, PRE 05], the method and uniformity of mixing the fuel and oxidizer nanopowders [PRE 05, SCH 05, BAH 14].

3.4.2. *Optical temperature measurement: spectroscopy*

Optical temperature measurements were also attempted in [KWO 03b] and [MOO 04]. Generally, qualitative differences in burning rate were observed with the apparent flame speeds using a high-speed camera. From these studies, burning rates for the nanothermites, with the same compositions, vary over a broad range (from 0.1 to 1,000 m/s) and appear to be most affected by the compaction of the sample. Apparent flame propagation measurements, often showing supersonic values of the flame speeds for some kind of nanothermites, are difficult to interpret as they change as a function of the sample heating and/or as the sample is compressed. The flame propagation remains poorly described in nanothermite powders and is currently the subject of intense research. At the moment, nanothermites are considered to deflagrate and not detonate.

3.4.3. *Photodiodes*

Photodiodes with collimated inputs were also used to register flame propagation instants to obtain a more accurate measurement of the flame speed [MOO 07]. Poor reproducibility between the combustion rate measurements in open tray experiments was often observed. The setup could be improved by installing a periodic sequence of baffles with small openings along the channel in which the sample is placed and ignited [WAL 07]. The baffles were meant to minimize the effect of removal or suspension of unignited free powder as a result of the strong convective flows produced by the propagating flames. Improved reproducibility was indeed achieved; however, the measured flame velocities decreased substantially and the mechanism of flame propagation remained unclear, as previously mentioned.

3.4.4. *Confined combustion tests*

Further modifications of the experimental methodology have included using cylindrical tubes in which the sample was packed and ignited [SON 07b, SAN 07, BOC 05, APP 07]. Usually, the tubes are equipped with multiple side openings for pressure transducers and fiber-optic cables feeding the flame emission signal to photodiodes. A schematic of the setup is given in Figure 3.19 and an image of an open-ended transparent acrylic tubing developed in [BOC 05] for visual inspection of the combustion front given in Figure 3.20. As the ends of the tube are typically open, such confinement of the combustion materials is only partial.

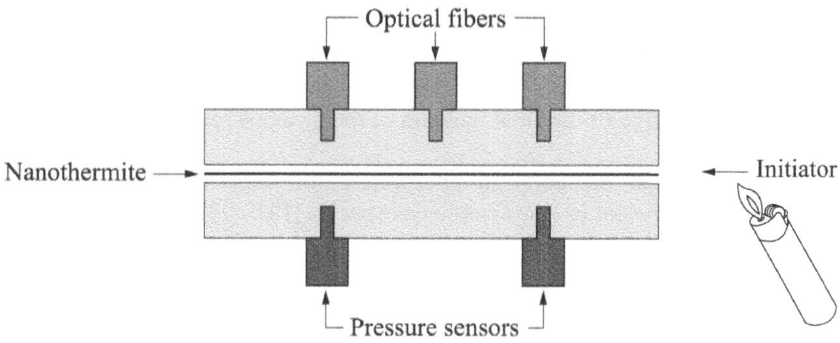

Figure 3.19. *Schematic of confined combustion test setup*

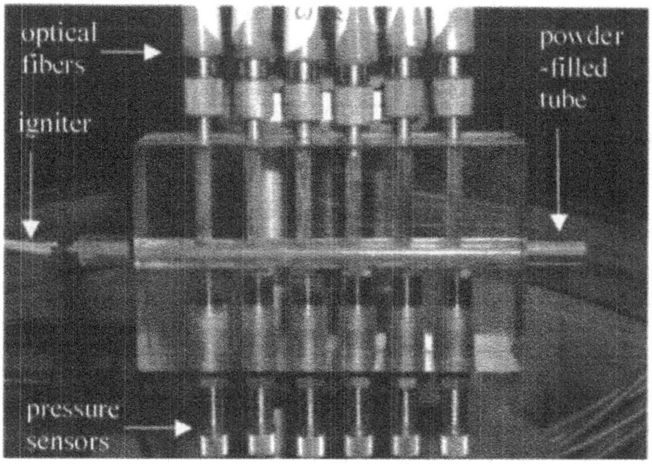

Figure 3.20. *Photograph of the instrumented tube setup as in [BOC 05] (Copyright 2005 American Institute of Physics)*

Another experimental technique used for the assessment of the nanothermite mixture combustion performance is based on a pressure measurement in a confined chamber, also called pressure cell [MOO 07, PUS 07, PUS 06, PRA 05, PER 07a]. Vessels of different sizes can be used and combined with different igniters[2] and pressure transducers. In this approach, the experiment can be considered as nearly adiabatic and the pressure increase can be related to the energy produced during the

2 Method or device able to heat the reactive mixture until its ignition temperature.

nanothermite combustion [PER 07a]. Both maximum pressures achieved and the rates of pressure rise are of interest. In addition, the final pressure in the chamber (after the reaction is completed and the products are cooled) is of interest as indicative of the final makeup of the gaseous reaction products.

One substantial difficulty in interpreting the experimental data is the production of transient or semi-stable gas species [PUS 07].

3.5. Ignition tests

Ignition of nanothermites can be accomplished by various phenomenons and techniques, including thermal, chemical, optical or mechanical impulses. Each involves a unique reaction mechanism, and, therefore, different nanothermite combustion behaviors. Initiation for impact-induced chemical reactions is linked to the degree of short time scale mixing and has been demonstrated to rely primarily on the intrinsic mechanical properties of the nanothermite constituents, whereby the material first undergoes a deformation and flow or fracture and dispersion prior to ignition. The onset can be correlated with the impact energy or stress.

Thermally initiated reactions using thermal analysis equipment, on the other hand, involve a thermochemical mechanism. This includes liquid-phase reactions based on evidence of observation of localized melt regions (called hot spots) in recovered partially reacted materials [VEC 94, STR 94] all of which conducted under low heating configuration. Heating rates normally fall in the range of $10–10^3$ °C/s which are several orders of magnitude slower than impact-induced reactions.

Thermal sources can be produced by Joule heating, radiation, laser or chemical reaction. A few studies investigated very rapid ignition regime where nanothermite heating rate exceeds 10^6 °C/s. In the following subsections, the different initiation conditions studied in the literature are reviewed.

3.5.1. *Impact ignition*

A study on the kinetics of impact-ignited chemical reactions in Al/Fe_2O_3 nanothermites prepared using ultrasonic mixing in hexane was conducted by Thadhani *et al.* in [CHE 10]. All samples were prepared at a stochiometric equivalence ratio of 1 ($\emptyset = 1$) and different sizes of reactants (Al and Fe_2O_3)

were tested. Final mixtures were compacted to reach a density of 70% TMD. Impact ignition of the thermite mixtures was performed using a compressed-gas gun producing an impact velocity ranging from 200 to 500 m/s.

Results of impact ignition of Al/Fe_2O_3 reactions at velocities up to 400 m/s reveal that the reactant particle size plays a more dominant role in the case of such type of ignition. It was found that a micrometer-sized thermite powder mixture reacts at a lower impact energy threshold than the mixtures of nanosized powders prepared by the same mixing technique. The difference in the reaction threshold is associated with the higher localized strain resulting from fewer interparticle contacts in micrometer-sized powders in comparison to a more distributed strain achieved during impact ignition of nanothermite pellet.

3.5.2. *High-rate heating (10^6–10^7 °C/s)*

The nanothermites can be heated rapidly with powerful heat sources, such as a detonation front produced by an ignition primer or as a fireball generated by a high explosive. In the latter case, heating rates are of the order of 10^6–10^7 °C/s. The generation of comparable heating rates while enabling controllable and reproducible conditions and accurate ignition detection is very difficult in laboratories. However, some studies use laser ignition with nanosecond and picosecond laser [YAN 02] pulses to produce heating rates in the order of 10^6–10^9 C/s [YAN 03]. Other works, such as [CHO 10], use thin platinum wires (76 μm in diameter) enrobed with the nanothermite to be ignited. The nanothermite is rapidly Joule-heated by a tunable voltage pulse generator. Such experimental results are well suited for studying kinetics of the ensuing vapor-phase reactions, which are, in most cases, irrelevant for heterogeneously driven ignition.

3.5.3. *Low and uniform heating (10–100 °C/s)*

The uniform heating of nanothermite mixtures is readily achieved in thermoanalytical experiments with very low heating rates.

Lasers also offer the capability of adjust and control energy transfer to nanothermite mixtures with a wide range of power. Laser ignition thus permits to set the heating rates of interest [GRA 04, DIM 89, KUO 93].

Granier *et al.* [GRA 04] studied the ignition time of slightly fuel-rich Al/MoO_3 composites (Ø = 1.2) prepared by ultrasonic mixing with a 50 W CO_2 laser. They compared the ignition of nanometer-size particles (diameters of 108, 39.2 and 29.9 nm) and micrometer-size particles (20 µm). Ignition delay corresponds to the time between the laser onset and the first detection of optical emission by the nanothermite pellets. Often, however, a number of issues make the interpretation of laser ignition experiments difficult. The most significant issues for nanothermites are the consistency in the sample preparation and differences in efficiency of absorption of the laser energy by different material components (e.g. metals vs. oxides or particles of different sizes present in the material). A different experimental approach was used to assess the scattering and absorption efficiency of Al nanoparticles and MoO_3 nanopowders in [BEG 07]. It was found that a dense nanopowder of MoO_3 scatters most of the incident light, while approximately two-thirds of the incident light are absorbed in a similarly packed Al nanopowder. The experimental approach that was used in [BEG 07] required the preparation of a slab with a moderate optical thickness, which was difficult for Al nanopowder. Additional measurements quantifying optical properties are needed for a broader range of nanoenergetic materials and incorporation of such data for interpretation of the laser ignition experiments is desirable.

3.6. Electrostatic discharge (ESD) sensitivity tests

Unintentional ignition of nanothermite is possible when static electricity is discharged into it. The most common form of ESD is the spark, which occurs when the static electric field strength exceeds approximately 4–30 kV/cm (the dielectric field strength of air is 30 kV/cm).

While there is a significant amount of literature for the ESD ignition sensitivity of high explosives, there is significantly less research on ESD ignition sensitivity of thermites. Recently, a few nanothermite mixtures were tested for ESD sensitivity using an apparatus as illustrated in Figure 3.21. Minimum ignition threshold energies were determined to be <1, 50 and 1 mJ for the Al/Bi_2O_3, Al/CuO and Al/Fe_2O_3, respectively.

Another study investigated three different sizes of Bi_2O_3 powder. The first powder has an average particle diameter of 2.4 µm, the second powder has an average particle diameter of 416 nm and the third powder has an average particle diameter of 80 nm. Each oxide is combined with 80 nm Al nanopowder using ultrasonic mixing and under stochiometric ratio (Ø = 1).

Results are given in Table 3.5. They clearly show the effect of decreasing oxide particle size in nanothermites susceptibility to ESD: decreasing oxide size lowers the ESD energy provoking the nanothermite ignition.

Average particle diameter (nm)	ESD Ignition sensitivity (μJ)
2,400	2
416	0.125
40	0.075

Table 3.5. *Effect of Bi_2O_3 average particle diameter on minimum ESD ignition energy of the Al/Bi_2O_3 nanothermite [KUO 08]*

Analysis of the ESD discharge testing shows that most mixtures consistently ignite at energies as low as 118 μJ, approximately 170 times less than the human energy storage capacity.

Nellums *et al.* [NEL 13] found that Al/Bi_2O_3 nanothermite mixtures prepared in resonant mixer in the polar solvent N,N-dimethylformamide (DMF) at high solids loadings (typically 30 vol%[3]) are five orders of magnitude less sensitive to ESD than dry Al/Bi_2O_3 powders.

Foley *et al.* [FOL 07] used Al/CuO nanothermite with Viton A in varying concentrations. They showed that the threshold ESD ignition energy is increased when Viton A is added to Al/CuO. One advanced explanation is that Viton A may effectively increase the mixture's electrical resistivity. They reasoned that a higher mixture electrical resistivity would lead to an increased minimum ignition energy threshold (increase in energy sensitivity). Beloni *et al.* [BEL 10] studied ESD ignition delay of micrometer-sized aluminum to characterize the sensitivity of these metal fuel powders to a discharged spark. Two different sizes of Al particles (~4 and 12 μm as nominal diameters) were tested and a linear correlation was found between burn time and Joule energy. Results showed that as the spark discharge energy increases, the delay time for ignition and burn time decreases [BEL 09]. They concluded that individual particles are heated by the spark and then begin heating and oxidizing the rest of the powder through Joule heating.

Weir *et al.* [WEI 13a] explored the relation between electrical conductivity and ESD sensitivity in terms of a discharged spark into different

3 Vol%: percentage by volume.

kinds of Al-based micron-size thermites: Al/CuO, Al/MoO_3, Al/Fe_2O_3, Al/I_2O_5 and Al/Bi_2O_3.

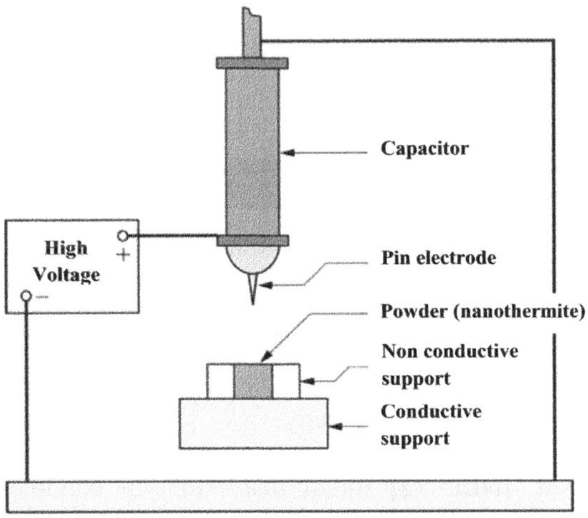

Figure 3.21. *Schematics of the ESD test apparatus*

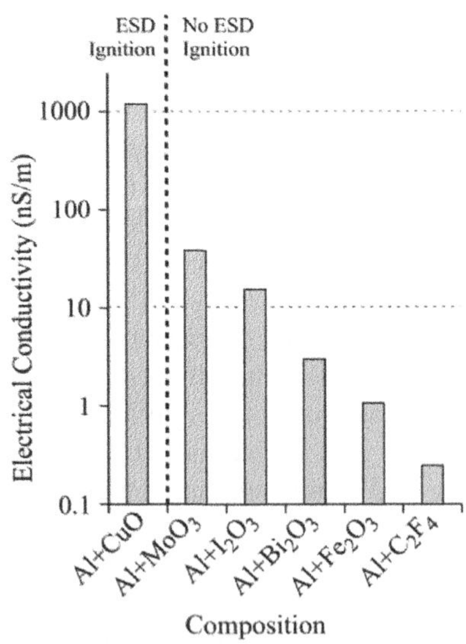

Figure 3.22. *Electrical conductivity of different nanothermites and ESD ignition under 10 kV [WEI 13a] (Copyright 2013 Elsevier)*

The only nanothermite mixture that ignited using ESD is Al/CuO. Figure 3.23 implies that there may exist a power and resistance threshold above or below which triggers ESD ignition. Further investigation is needed to understand more completely the relationship between electrical conductivity and ESD sensitivity. As the Al particles are from the same manufacturer for all nanothermites, ESD ignition is also a strong function of the oxide nature. Similarly, Weir *et al.* [WEI 13b] showed that as Al particle size is reduced, electrical conductance increases dramatically, as does ESD ignition sensitivity. Overall, electrical conductance is shown to increase linearly with increasing Al surface area to volume ratio and the alumina passivation shell surrounding Al core particles plays a significant role in enhancing ignition sensitivity.

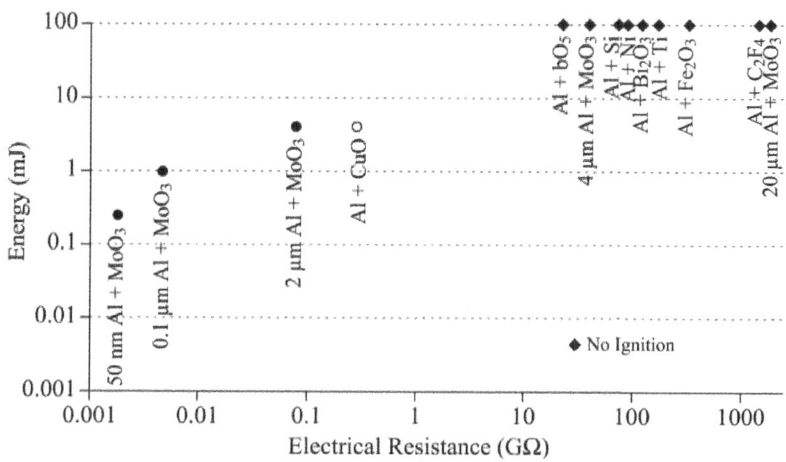

Figure 3.23. *Energy as a function of electrical resistance. Circular symbols correspond to the minimum ignition energy and the star symbols correspond to the maximum energy for which ignition did not occur [WEI 13b] (Copyright 2013 Elsevier)*

It is also now well known that reducing the nanothermite ESD sensitivity can be done just by incorporating a small amount of carbon. Less than 1 weight% is enough to reduce ESD sensitivity without penalizing reactivity.

4

Other Reactive Nanomaterials and Nanothermite Systems

In addition to aluminum (Al) and oxide powder mixing, other technological routes were explored to build nanoenergetic materials and integrate them into reactive systems. Most of the works are based on modern technologies such as sol–gel chemistry, vapor deposition method, cold spray and arrested reactive milling (ARM) to build the reactive material atoms-by-atom and thus accurately control the structures, intimacy and distribution of the oxidizer and fuel simultaneously. Many of these new technological methods also allow to integrate the nanoenergetic material directly on a functional device.

In this chapter, a review of these different technological routes is presented and the main characteristics and performances of the reactive materials obtained are analyzed. The chapter is divided into six sections: the sol–gel materials (section 4.1), multilayered foils (section 4.2), arrested milling dense materials (section 4.3), core–shell structures (section 4.4), reactive porous silicon (section 4.5) and others (section 4.6).

4.1. Sol–gel materials

An alternative to mechanical mixing of nanopowders is sol–gel processing proposed in a few papers [GAS 01a, GAS 01b, TIL 01]. The sol–gel method is a bulk chemical technique used for the synthesis of materials with nanometric dimensions.

Historically, sol–gel chemistry has employed the use of metal alkoxide precursors that readily undergo catalyzed hydrolysis and condensation to form a gel of metal oxide particles with nanoscale dimensions (1–100 nm). Researchers at Lawrence Livermore National Laboratory synthesize porous monoliths and powders of nano-sized transition metal oxides such as Fe_2O_3 or Cr_2O_3 combined with metals such as Al, Mg and Zr. Hydrated salts of metals such as $Fe(NO_3)_3.9\ H_2O$ or $Cr(NO_3)_3.9\ H_2O$ or $FeCl_3.\ 6\ H_2O$ serve as precursors, and epoxide serves as a gelation agent. Dissolved in solution, these precursors undergo hydrolysis and condensation to form a stable sol (suspension made of metal nanoparticles in solution). This process is initiated by changing the pH, temperature, or ionic strength of the solution or through the addition of a catalyst or gelling agent. Through manipulation of these conditions, the sol can be further linked through condensation of surface groups on the particle surfaces to form a gel. The gel is a rigid three-dimensional structure that has a nanostructured framework and pores with dimensions of 2–100 nm in diameter. The resulting wet gel can be dried under atmospheric conditions to yield a xerogel. A xerogel is a porous material with high surface area (50–500 m^2/g) and a moderate density (typically 20–75% of that of the bulk). Alternatively, supercritical extraction with CO_2 of the solvent in the wet gel leads to an aerogel. An aerogel is a highly porous material with a high surface area (100–1,000 m^2/g) and low density (typically only 1–25% of that of the bulk). The process is carried out preferably in a polar protic solvent.

Figure 4.1 summarizes the main steps of sol–gel chemistry and Figure 4.2 presents a photo of one final nanothermite material obtained by sol–gel chemistry.

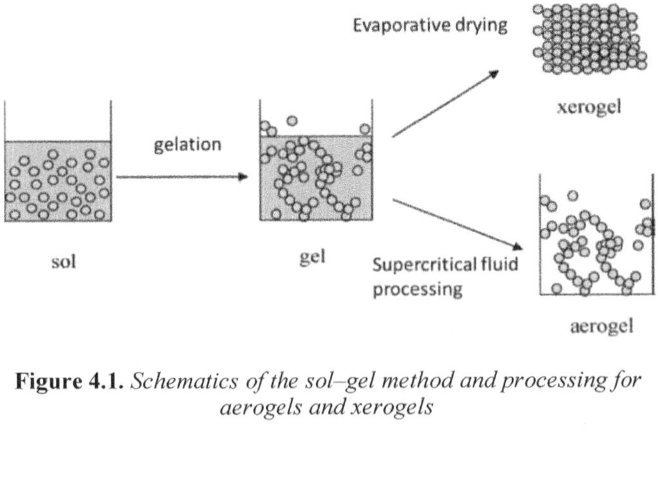

Figure 4.1. *Schematics of the sol–gel method and processing for aerogels and xerogels*

Figure 4.2. *Photo of one chromia aerogel pellet [GAS 01a] (Copyright 2001 Elsevier)*

Figure 4.3 depicts the typical transmission electron microscope (TEM) images of the aerogel of Al/Fe_2O_3 thermite synthesized by the Lawrence Livermore National Laboratory [TIL 01]. We can see that the Al particles (30 nm in diameter) are immobilized within the Fe_2O_3 skeleton ensuring good contact. However, unavoidable organic impurities were also detected as a result of residual solvent which drastically penalizes the final material energetic performances. Indeed, these organic impurities act as heat sinks during flame propagation. In [PLA 05], Plantier *et al.* showed that heating the oxidizers helps to reduce the quantities of impurities in the aerogel and xerogel leading to a significant increase in burning rates. The nanothermites with aerogel showed the largest increase in burning rate from ~10 to over 900 m/s.

Ta/WO_3 reactive materials were also prepared by sol–gel and compared with conventional powder mixture, and both samples were consolidated using the technique of spark plasma sintering. The sol–gel composite released heat that was approximately 30–35% higher than that of the powder mixture. This result is attributed to the presence of carbon in the sol–gel composite. Additional ignition tests show that the sol–gel Ta/WO_3 reactive material obtained is insensitive to friction, spark and impact ignition.

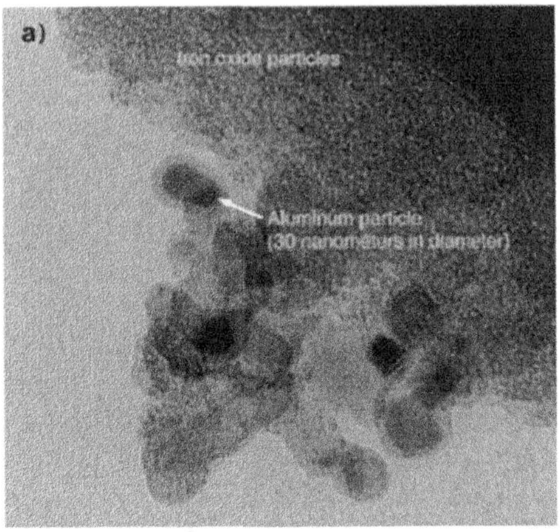

Figure 4.3. *Transmission Electron Microscopy image of Al/$Fe_2 0_3$ aerogel [TIL 01] (Copyright 2001 Elsevier)*

Sol–gel approach is attractive for nanothermite synthesis as it naturally generates a very intimate mixing between the fuel and oxide components, it is done at room temperature, and is low-cost and easy. The disadvantages include high porosity of the final thermite (the surface area can reach 400 m^2/g). Other disadvantages are restrictions on the types of materials that can be gelled, and a difficulty in scaling up the production rate – mainly because of the need to introduce the metallic nanopowders.

4.2. Reactive multilayered foils

Another distinct class of nanoenergetic materials is represented by reactive nanolaminates, also called reactive multilayered nanofoils or foils, wherein two or more different materials are stacked on top of each other. These multilayered structures can sustain a self-propagating exothermic reaction once it is initiated locally thermally, optically or by a mechanical shock. Figures 4.4 and 4.5 show the cross-sectional images of Al/CuO and Al/Ni multilayered foils prepared by physical vapor deposition [PET 10b] and mechanical rolling [STO 14, STO 13, ADA 15].

Research on reactive multilayered foils dates back to the early 1990s. One of the first manifestations of multilayered reactive foil was presented in 1990 by Thompson *et al.* [MA 90] in the form of Al/Ni films with individual layer thicknesses between 60 and 300 nm (alternatively deposited by e-beam evaporation at ~ 10^{-7} Torr). This approach, protected in a patent [BAR 96], has been followed by extensive research in the USA and to a lesser extent in Europe. Studies of multilayered reactive materials have been motivated from both a fundamental science perspective and by potential applications. The former has particularly aimed at taking advantage of the fact that these materials, as model systems, offer a unique setting for analyzing phase transformations under different heating conditions [BLO 03, EDE 94, BAR 97, TRE 10, KIM 11, TRE 08, FAD 10, ZHA 07b, BAR 11, WEI 11a]. It also offers an ideal structure for the development of analytical, empirical or numerical models to understand metal–oxide or metal–metal combustion. As application, various areas including joining, brazing, sealing and ignition of secondary reactions have been explored. This will be detailed later in Chapter 6.

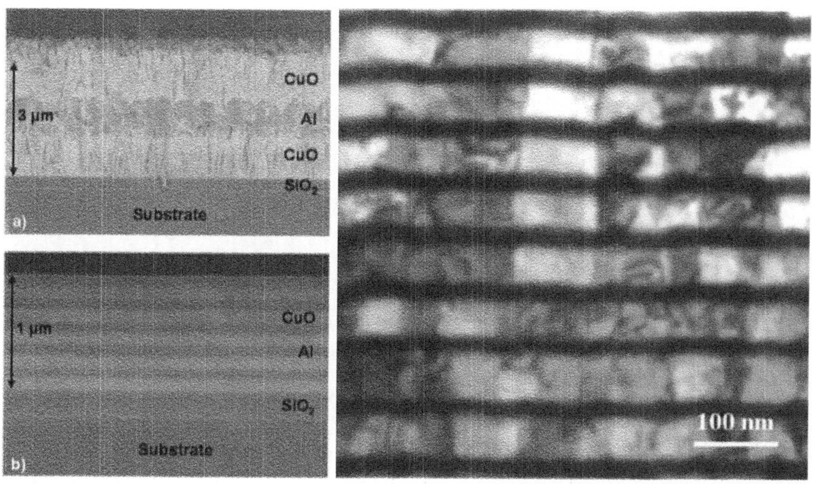

Figure 4.4. *Left: High Resolution Transmission Electron Microscopy images of left CuO/Al multilayers. a) Three layers of CuO (1 µm)/Al (1 µm)/CuO (1 µm); b) 10 layers of CuO (100 nm)/Al (100 nm) [PET 10b]. Right: Ni/3Al [BOE 10] (Al/Ni NanoFoil©)*

4.2.1. *Bimetallic multilayered foils*

4.2.1.1. *Method of fabrication*

Most of the research efforts have proposed bimetallic films with individual layer thicknesses between 5 and 500 nm. Among bimetallic, Al/Ni is the most studied but some teams have also considered Ni/Si, Ti/Si, Va/Si, Ti/Al Ti/Al/B, Al/Pt, Al/Ni/Fe, etc.

They can be obtained either by the physical vapor deposition (PVD) method [CLE 90, MA 90, REI 99, GAV 00, WAN 04a, GAC 05, ADA 06, DYE 94] or by a mechanical process [BAT 99, SIE 01, HEB 04, QIU 07]. PVD methods provide excellent control over layering and, therefore, reactant spacing. However, PVD is typically a time-consuming and fairly expensive process, limiting the quantity of reactive material that can be fabricated using this technique. In contrast, mechanical rolling permits average bilayer spacing in the micrometer scale with a broad distribution of bilayer spacing, but is preferred in some applications due to the much lower production costs and the ability to fabricate large quantities of materials.

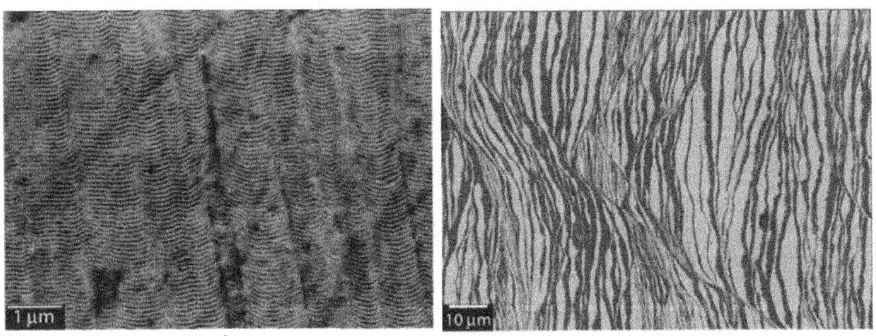

a) b)

Figure 4.5. *Scanning Electron Microscopy images of a) vapor-deposited and b) mechanically processed Al/Ni multilayer foils [KNE 09] (Copyright 2009 American Institute of Physics)*

4.2.1.1.1. Mechanical methods such as rolling

Battezzati *et al.* [BAT 99] proposed a cold rolling technique to create Al/Ni alterning foils followed by an annealing phase to form intermetallic with the goal of phase transformation investigation. Repeated low energy pressing or cold rolling of foils never ended in the formation of intermetallics, but in refined microstructures made up of intimately mixed elemental foils. The available observations show that the length scale of layer mixing may reach a few tens of nanometers. Intermetallics are then formed by annealing the foils at temperatures between 200 and 600°C.

In [STO 14], Stover *et al.* described a mechanical method for creating reactive laminates using the rolling process. Bulk reactive materials are created by cold rolling stacks of alternating sheets of nickel and aluminum into foils with bilayer thicknesses in the micrometer scale range. The rolled foils are then ground into laminate powders and sieved. This method permits us to produce reactive multilayered foils with variable sizes. A schematic of the cold rolling process is given in Figure 4.6.

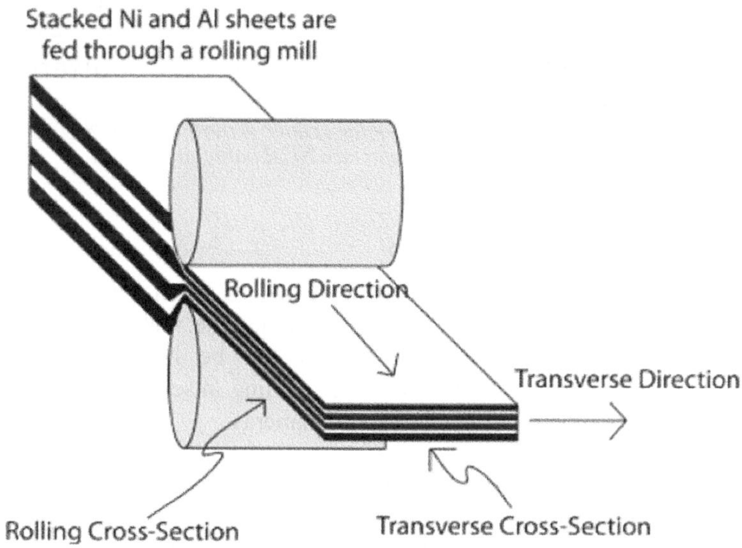

Figure 4.6. *Schematics of the cold rolling process proposed in [STO 14] (Copyright 2014 Springer)*

4.2.1.1.2. Physical vapor deposition methods

PVDs include sputtering or electron beam evaporation, the most used being magnetron sputtering. The main principle depicted in Figure 4.7 is the alternative deposition of each metal on a substrate. The process parameters are target power, gas flow, chamber geometry and target-substrate distance.

After the deposition, the foils can be released from the substrate and characterized in combustion.

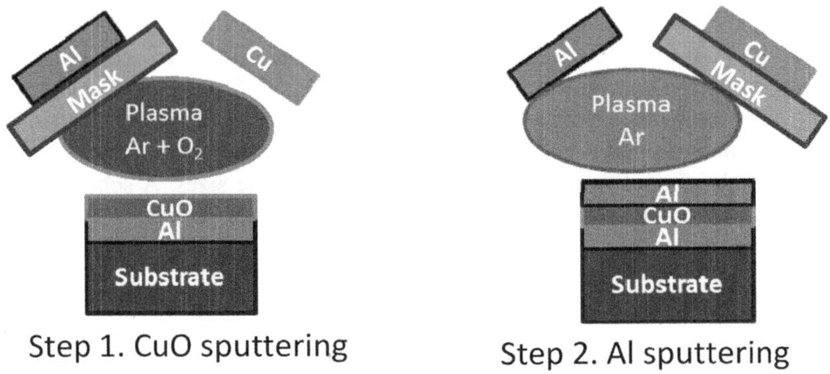

Figure 4.7. *Schematic view of the sputter deposition process proposed to manufacture Ni/Al nanofoils*

4.2.1.2. *Properties and performances*

Reactivity in these multilayered materials is often characterized by measuring heats of reaction and flame velocities, i.e. velocity of the self-propagating reactions. Sustained reactions can be initiated in reactive multilayers via electric or thermal stimuli, resulting in self-propagating flame fronts that can be monitored by high-speed cameras.

Measured flame velocities of bimetallic multilayers are in the range of 1–10 m/s. When the bilayer thickness (or spacing) decreases down to tens of nanometers, self-propagating flame velocities rapidly increase to reach the maximum value. For instance, for the Ni/Al system, flame velocity exceeding 10 m/s is observed [KNE 09, BRA 12a, BRA 12b, BRA 12c].

Weihs *et al.* have analytically and numerically developed models able to predict the average self-propagating reaction rate, V^{avg}, as a function of the bilayer spacing (~4δ) and the width of the premix region (4*w*) as drawn in Figure 4.8.

As discussed in [MAN 97], the partially mixed region results in a loss of thermodynamic driving potential, due to premixing during the deposition process (see section 4.1.2). V^{avg} increases with decreasing δ, as long as δ remains substantially larger than the premix width *w*. This trend reflects the drop in the mean diffusion distances as δ is decreased.

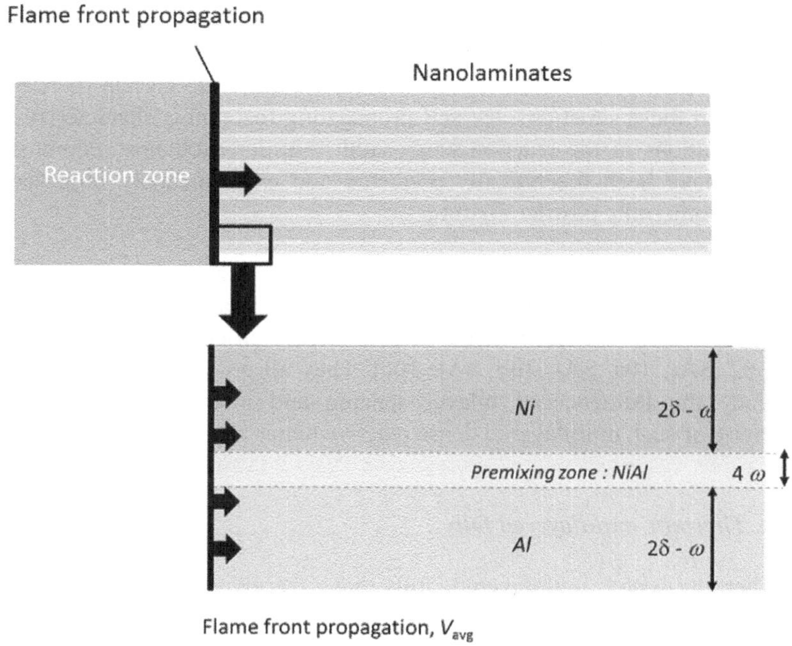

Figure 4.8. *Schematic of an unreacted Ni/Al bilayer separated by a thin NiAl premixed region*

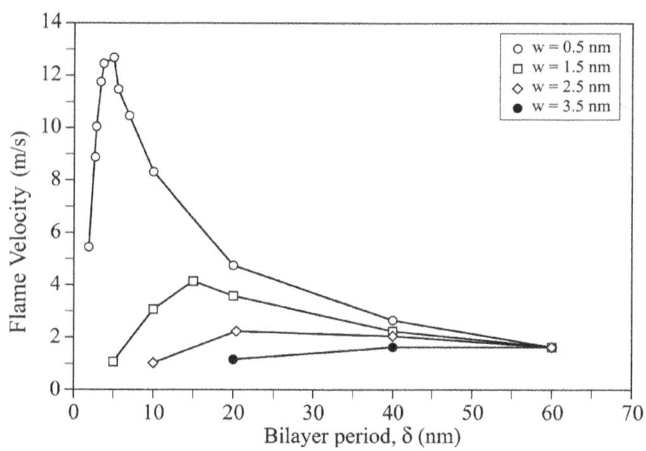

Figure 4.9. *Average self-propagating reaction velocity as a function of δ for Ni/Al foils with different premixing widths, ω [MAN 97] (Copyright 1997 American Institute of Physics)*

However, as δ is further decreased, velocity drops rapidly to zero. Numerical models [JAY 98a, JAY 98b] have also shown that in addition to steady propagation, oscillatory combustion occurs in certain parameter regimes. In these situations, the self-propagating reaction is characterized by large variations in the reaction zone width. Similar oscillatory combustion regimes have been theoretically predicted and experimentally observed in dense nanothermite powder mixture.

Subsequent computational studies have aimed at incorporating more elaborate models to account for property variation, phase change effects, heat losses, as well as the dependence of heat release on composition [JAY 99, BES 02, SAL 10a, SAL 10b, SAL 10c]. They all predict with more or less accuracy the influence of bilayer spacing and intermixing zone in the reactivity of such multilayered dense reactive materials.

4.2.2. *Thermite multilayered foils*

Thermite-based multilayered foils have received comparatively less attention than bimetallic foils, but they are significantly interesting for some applications because the thermite reaction is more energetic than the intermetallic reaction.

Bimetallic	Theoretical enthalpy of reaction (kJ/g)
Al/Co	1.3
Al/Ni	1.4
Al/Ti	1.0
Al/Fe	0.9
Al/Pt	0.9
Al/Cu	0.5

Table 4.1. *Enthalpy of reaction for a few bimetallics (to compare with main thermite reaction see Table 3.1)*

The Al/CuO multilayered foils were selected as the first research object [BES 02]. Al/CuO multilayered foils with different bilayer spacings have been deposited and their reactions were compared in several papers [BAH 14, PET 10b, MAN 10].

4.2.2.1. *Method of fabrication*

They are obtained by PVD methods providing excellent control of the layering and purity of the final material since the process is done at very low pressure.

Typically, as described in [BAH 14], Al/CuO nanolaminates are direct current (DC) magnetron sputter deposited on silicon wafers from Al and Cu targets. During the sputter deposition, the temperature of the substrate is regulated at 10°C and the base pressure of the chamber is less than 10^{-5} Pa. During deposition, crystalline Cu_4O_3 and CuO structures could be produced, depending on the oxygen partial pressure during deposition. Many process parameters have to be accurately set to obtain pure and well-controlled individual layers, such as the oxygen to argon flow rate, the working pressure, the target power, as well as the bias voltage.

For example, hereafter, a deposition process enables us to achieve predominantly uniform copper oxide with a Cu/O ratio of 1. CuO is deposited under argon and oxygen plasma, from a Cu target with an oxygen partial pressure of 0.13 Pa and oxygen to argon flow rate of 0.25. Aluminum thin film is deposited from an Al target under argon plasma.

To produce multilayered foils, the substrate (typically silicon) is successively exposed to the deposition of CuO first, and then Al without changing the chamber. The process is depicted in Figure 4.10. A photo of a typical chamber is given in Figure 4.11.

After the CuO layer deposition, the chamber is automatically pumped out to remove oxygen in order to prevent aluminum oxidation during the Al deposition cycle. Each layer thickness can be set between 25 nm and 2 μm with an accuracy of 5 nm.

Note that Al/CuO multilayered systems are difficult to be sputter-deposited when bilayer spacing decreases down to 50 nm because Al+CuO is highly reactive and heat or charges generated by the sputtering process can be sufficient to locally ignite the exothermic reaction that can damage the sputter deposition equipment. Figure 4.11 illustrates wafers ignited during sputter deposition process.

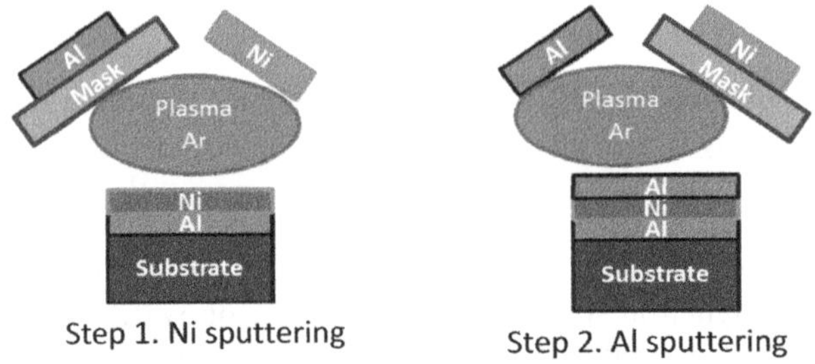

Figure 4.10. *Schematic of Al/CuO multilayered film sputtering process described in [BAH 14]*

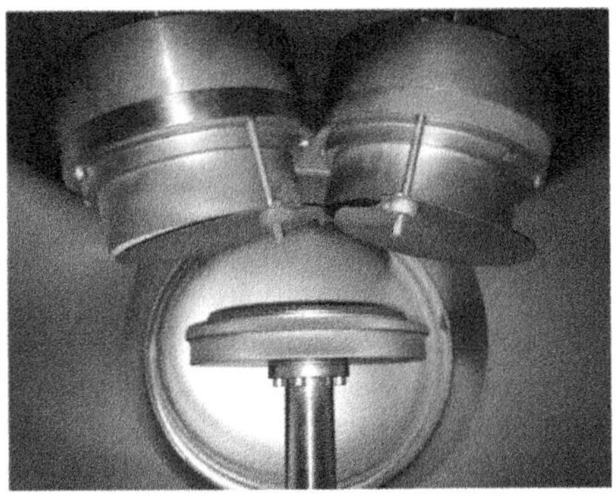

Figure 4.11. *Photo of the inside of the sputtering chamber: top left is the Al target, top right is the Cu Target and the substrate is seen at the bottom of the photo*

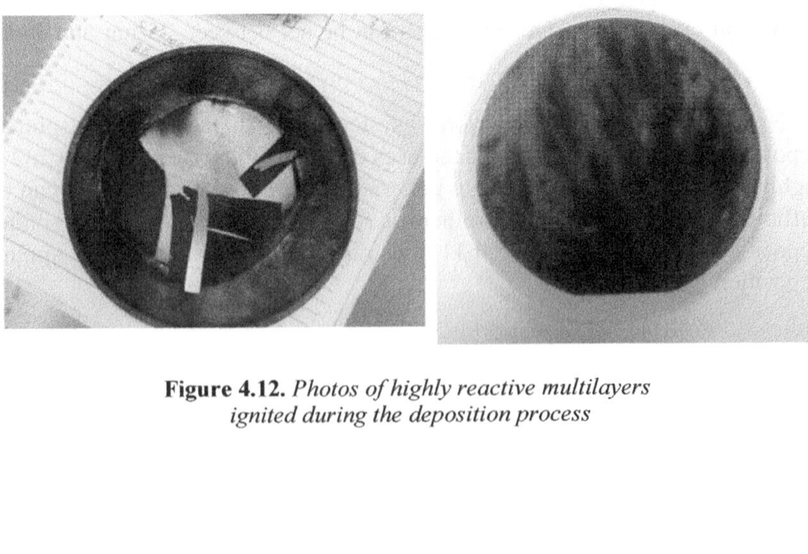

Figure 4.12. *Photos of highly reactive multilayers ignited during the deposition process*

4.2.2.2. *Properties and performances*

When the Al+CuO layer thicknesses are scaled down to 25 nm, the material becomes unstable and the film can ignite spontaneously at room temperature, underscoring the key role of the interfacial region (nature, defects and homogeneity). The interfacial region is the layer between the Al and CuO layers, being of a few nm in thickness.

A comprehensive and detailed study of the nature and role of interfacial layers in reactive properties of Al/CuO nanolaminates has been recently published [KWO 13]. The most important issue in sputtering of thermite multilayers is that during material sputter deposition, between Al and oxide layers, an intermixing naturally occurs to form the so-called diffusion barrier layer. This interface occurs spontaneously between Al and CuO and its thickness and nature affects the reaction kinetics [KWO 13].

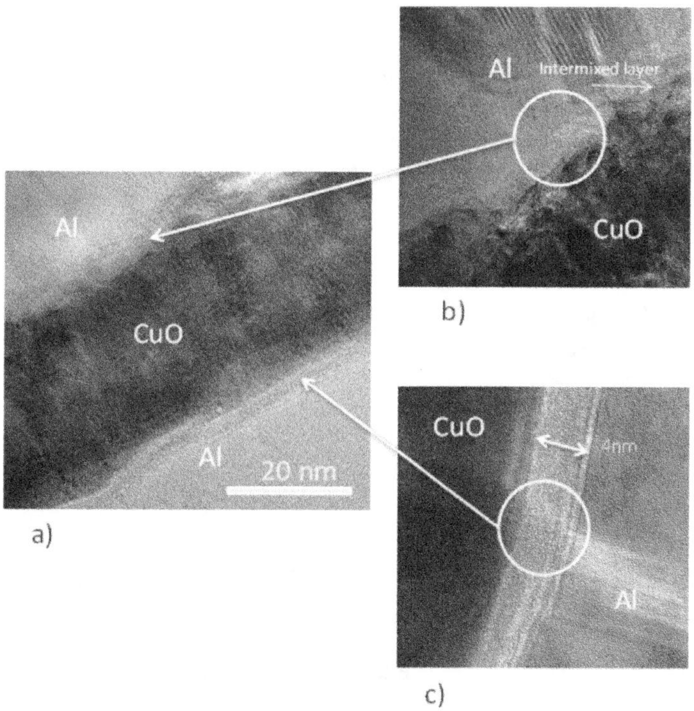

Figure 4.13. *High-resolution transmission electron microscopy images of a) the Al/CuO nanolaminate showing the columnar grain structure of the CuO layer (dark zone) and the grains in the Al layer (lighter contrast); b) sputter deposited of CuO on Al and c) sputter-deposited of Al on CuO [KWO 13]*

The HR-TEM images in Figure 4.13 show interfaces between aluminum and copper oxide, both sputter deposited. Figure 4.12(b) corresponds to CuO deposition on an Al-50 nm layer: the interface is regular in thickness with an average of 4 nm (~ between 3 and 5 nm) and amorphous. It is likely that an Al_xO_y layer is formed when O_2 plasma is generated inside the chamber just before starting the Cu sputtering; in other words, the virgin Al surface is bombarded with O atoms or O_2 molecules first (before being sputtered) and then with sputtered Cu from the Cu target. We also note that this interface formed when CuO is sputtered on Al is thicker than the native oxidation of Al.

The energetic properties of thermite multilayered materials are characterized by measuring heat (enthalpy) of reaction in differential scanning calorimetry (DSC) equipment and flame velocity after ignition via a thermal stimulus. Measured flame velocities are in the range of 2–90 m/s for thinnest bilayer spacing (one bilayer corresponds to Al+CuO layers). The plot of the self-propagation rate against the bilayer spacing (Figure 4.14) in stoichiometric condition (Ø = 1) shows that the self-propagation rate followed the same trend. It decreases from 90 m/s, for the thinner bilayer thickness (50 nm), down to 2 m/s for the thickest layering (1,500 nm). When bilayer thickness decreases below 50 nm, there is no sustained combustion. These obtained self-propagation rates are well below those reported for nanopowder mixing.

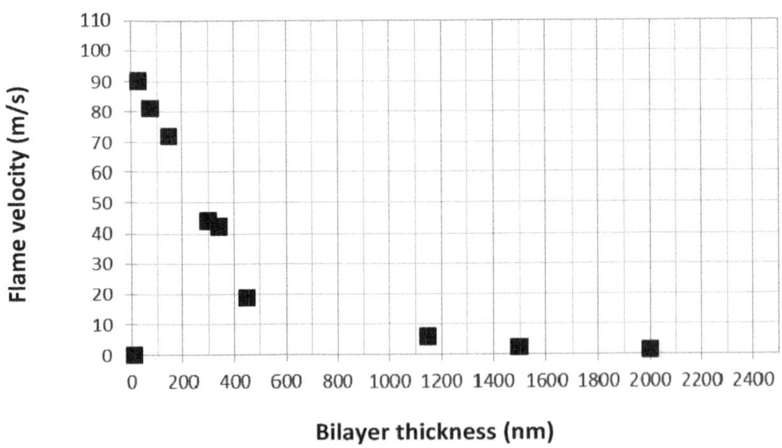

Figure 4.14. *Measured self-propagating flame front as a function of bilayer spacing for Al/CuO and sputter-deposited in stoichiometric ratio (Ø =1). Total foil thickness is of 2 μm*

Note that, compared to other multilayered systems such as Al/Ni systems, the Al/CuO multilayered systems are characterized by higher reactivity.

The effects of substrate material and thickness on the self-propagation rate of Al/CuO multilayered systems have also been examined, both experimentally and theoretically. Manesh *et al.* in [MAN 10] prepared Al/CuO multilayered systems having a total thickness of 3.2 μm by magnetron sputter deposition technique. Al layer thickness of 26 nm and CuO layer thickness of 54 nm were used to provide a bilayer of 80 nm. The thin film was deposited on substrates of various thicknesses and thermal conductivities: glass or silicon. Average self-propagation rate is 46 m/s on glass and reaction quenched on Si showing the importance of thermal environment.

4.2.3. *Summary*

Vapor deposition and rolling are the two explored methods to prepare multilayered reactive films and foils. Multilayer vapor deposition is generally versatile because almost all of the frequently used metals, metalloids and metal oxides can be prepared by choosing suitable deposition parameters, and it is convenient to control the thickness of the layers and the number of layers. By changing the thickness of the layers, the number of layers and/or changing the materials (Sb, Nb, Si, Al, Ni or some oxide) in the system, the self-sustained propagation rates are easily tunable [BAH 14, REI 99, GAV 00]. This perspective of tunability along with other attributes such as high volumetric energy density and the capability to integrate on a chip environmentally benign products without manipulating dangerous reactants or reactive Al nanopowders make the PVD process and material very safe and attractive for commercial products for wide applications, including environmentally clean primers, detonator, explosives, *in situ* welding and many others that will be discussed (see Chapter 6).

4.3. Dense reactive materials

Another category of nanoenergetic materials enters into dense reactive materials, i.e. featuring a theoretical maximum density (TMD) percentage greater than 70%. They can be obtained in two ways: either mechanically by high-energy ball milling (HEBM) or after a consolidation process under high gas pressure. Both methods are detailed hereafter and the material characteristics are briefly presented.

4.3.1. *Arrested reactive milling*

HEBM has been widely applied in the rapid combustion synthesis of materials and can effectively generate nanostructured energetic or thermite materials. ARM is a typical example of this methodology, and many groups of research have applied this method to produce nanothermite materials. The milling of metals and metal oxides represent the most used compositions, but boron and metals such as titanium, zirconium or hafnium, forming respective borides, represent another class of useful compositions. A comprehensive review has been provided by Dreizin *et al.* in [DRE 07] which is the major contributor in this technique.

A mixture of regular micrometer-sized fuel particles capable of reacting exothermically (metal and/or oxide powders) is mixed and ball milled. The powders are loaded at room temperature with milling balls. The milling is performed at a rotation speed of several hundred rounds per hour. The exothermic reaction is, therefore, initiated mechanically and becomes self-sustaining rapidly resulting in substantial increases in both the pressure and temperature within the milling vessel. Transient pressures can reach up to 5 GPa. The temperature is controlled by a cooling jacket filled with water or, if necessary, with liquid nitrogen. Concerning this last case, we speak of cryogenic milling.

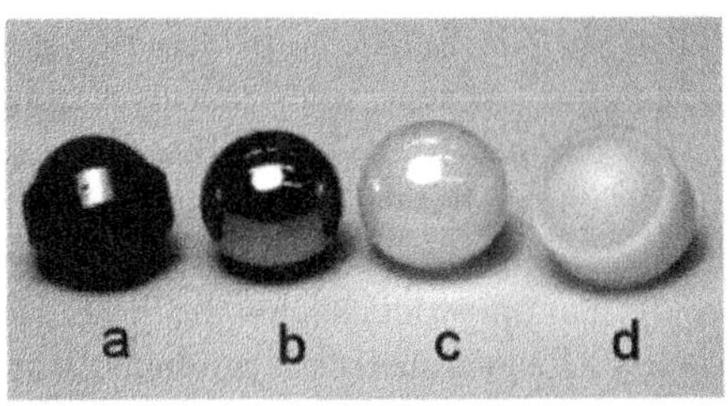

Figure 4.15. *Photos of milling balls used in the synthesis of the Al/MoO_3 reactive materials: a) tungsten carbide; b) steel; c) zirconia and d) alumina [UMB 06a] (Copyright 2006 Wiley)*

The milling process is then interrupted or arrested just before the self-sustaining reaction is mechanically triggered. The variable parameters include the initial powder size, the mass ratio of powder sample to that of the milling media (hexane), the processing temperature and the milling time. The adding of hexane permits us to suppress the initiation of the self-sustained regime. ARM leads to the formation of fully dense, micrometer-sized pellets with nanoscaled structural features. The final material is a three-dimensional composite of the initial powder near the theoretical maximum density.

Figures 4.16 and 4.17 give two examples of dense reactive materials obtained by ARM process.

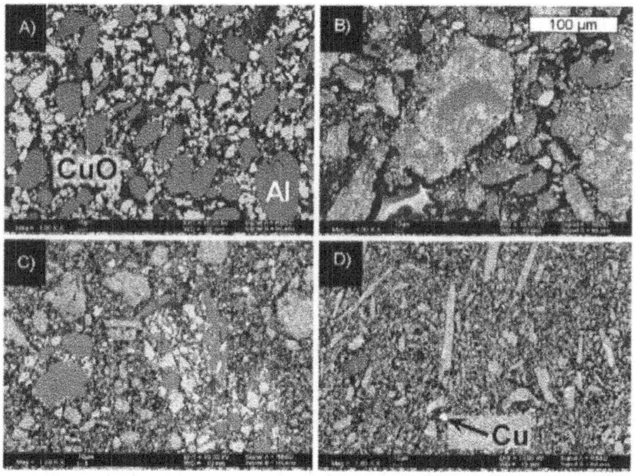

Figure 4.16. *SEM images of the cross-sectioned Al/CuO samples embedded in epoxy: a) starting material; b) c) and d) are the milled samples under different conditions (hexane quantities and milling times were variable) [UMB 06b] (Copyright 2006 Elsevier)*

Figure 4.17. *SEM images of the milled 2Al/MoO_3 with different hexane conditions (milling time was set at 60 min) [UMB 06a] (Copyright 2006 Elsevier)*

ARM processing is very flexible, versatile and inexpensive. It has been observed that essentially any combination of metallic and oxide powders can be processed by ARM to prepare a nanoenergetic material.

Below is the list of some nanomaterials prepared by ARM. It includes various thermites as well as bi-metal systems: Al/Fe_2O_3, Al/CuO, Al/MoO_3, Al/Bi_2O_3, Al/WO_3, Al/SrO_2, $Al/NaNO_3$, Mg/MoO_3, Mg/CuO, Si/ CuO, Si/MoO_3, Si/Bi_2O_3, Zr/MoO_3, Zr/CuO, Zr/ Bi_2O_3, $Si/NaNO_3$, 2B+Ti, 2B+Zr, Al+W, Al+Hf, Al+Mg, Al+Ti, Al+Zr, etc. In Europe, Rebholz *et al.* explored the low-energy ball milling process to produce dense pellets of bi-metallic as Al/Ni with a molar ratio of 1:3, as described in [HAD 10a] and [HAD 10b].

One important limitation of ARM processing is the inevitable presence of a fraction of reacted material in the final material, since during the processing, a reaction between the components is locally triggered mechanically but not self-sustained within the sample. As a result, the small quantities of the reaction products are redistributed within the sample that is continued to be ball milled. The presence of the partial reaction products can be minimized by adjusting the milling conditions and parameters [UMB 06a, UMB 08], but it cannot be avoided. Another important issue that needs to be considered when preparing nanoenergetic materials by ARM is the safety of the ball mill operation, specifically, the self-sustaining reaction between the reactive components must be absolutely prevented to avoid the damage to the processing equipment and facility.

Importantly, as in multilayered foils, the oxide (MoO_3 in the example of Figure 4.16) is fully embedded into the matrix of Al which is not the case in powder mixing process. Therefore, the entire Al/oxide interface area will participate in the exothermic heterogeneous reaction upon initiation triggering.

Detailed experimental studies of exothermic reactions in the ARM-prepared nanothermites showed that the reaction onset is observed at much lower temperatures than nanopowder mixing, between 127 and 227°C (value obtained by DSC measurements). This significant difference in the onset temperature for the exothermic reaction suggests that the reaction mechanism is different for the ARM-prepared thermites as compared to the powder mixing mixtures containing naturally passivated aluminum particles. Flame propagation measured in open channels was greatly lower (1 or 2 decades lower) than those of powder mixture which is a direct effect of density

(TMD% > 70%). In [SCH 05], the authors report a burning rate of 0.14 m/s obtained under atmospheric conditions and for a Al/Fe_2O_3 thermite. The propagation was generally faster and more uniform for the powder for which the milling was arrested just prior to the expected spontaneous reaction.

These results are totally in agreement with expectations: the reaction propagation mechanisms in fully dense nanomaterials are completely different than those in highly porous mixtures of nanopowders. In reactive material, pores play a critical role in promoting the pressure-driven reaction propagation, while a slower, thermal reaction propagation mechanism is expected for the fully dense materials.

For aluminum-rich Al/CuO ARM-prepared thermite, ignition in air occurs at 560–650°C, which is close to what obtained in Al/CuO nanofoils [BAH 14].

4.3.2. *Cold-spray consolidation*

Bacciochini *et al.* in [BAC 13] used a cold gas dynamic spray (called cold spray) process combined with ball milling to consolidate reactive materials and obtained highly dense material (up to 99% of the theoretical maximum density).

Results on Al/CuO-consolidated thermite show a relation between the sample's porosity level and the flame propagation velocity. The reduction of porosity tends to decrease the burning rate from approximately 14 m/s at 20% density down to 1.5 m/s at 100%. Once again, this indicates a transition from convection dominated energy transfer at low porosities to heat conduction dominated mechanism at vanishing porosities when gas percolation is not possible.

Figure 4.18 represents the cold spray process principle: a pressurized gas is heated before flowing through a converging–diverging nozzle creating a supersonic flow. The powder is injected in this flow, which is accelerated by the pressurized carrier gas (commonly He). Upon the impact on the substrate, particles deform and build up a deposit by intense local plastic deformation in the solid state. The compaction pressure ranged from 0 to 60 MPa resulting in average final densities ranging from 20 to 60% of the TMD.

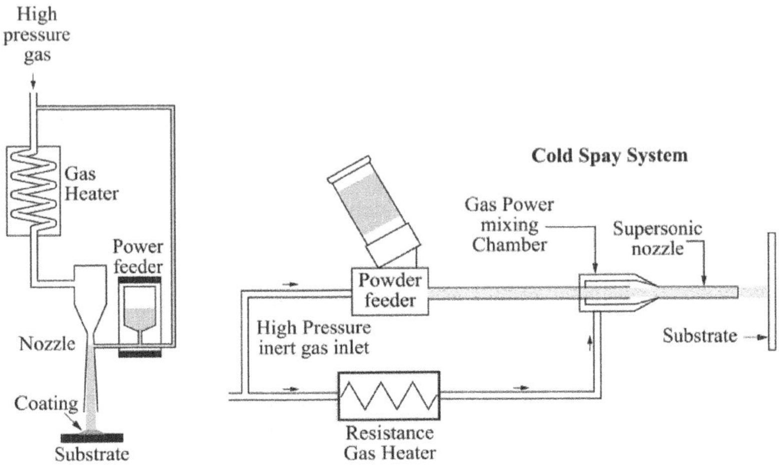

Figure 4.18. *Schematics of the cold gas dynamic spray process as in [BAC 13] (Copyright 2013 Elsevier)*

In [DEA 13], cold spray was used to synthesize two bi-metallic reactive materials from powders composed of mixed Ni/Al and mixed Ni/Al/MoO_3, under following conditions: the pressure is 2.1 MPa and working gas is He. The working temperature is ~178 °C.

The energetic properties of cold sprayed materials were compared with lower density axially pressed pellets of identical composition. Samples composed of a mechanical blend of Ni and Al powders have higher reaction rates at lower densities; with the near fully dense cold spray samples having the lowest burning rates.

Figure 4.19. *Photo of a Ni/Al energetic material produced by cold spray [DEA 13] (Copyright 2013 Elsevier)*

4.4. Core–shell structures

Core–shell structures are a class of materials which have properties intermediate between those of small, individual atoms or molecules and those of bulk.

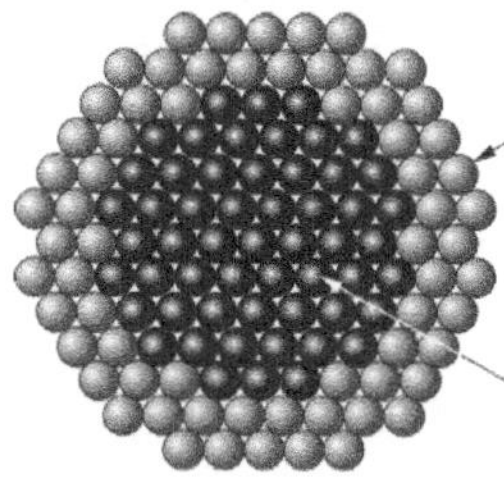

Figure 4.20. *Schematics of a core–shell structure*

First core–shell nanoenergetic structures have been imagined by Menon *et al.* in 2004. Fe_2O_3 nanowire arrays were partly embedded into thin aluminum films [MEN 04] as depicted in Figure 4.21. The process started with electrochemical anodization of Al foil to form nanoporous alumina templates. Then, Fe nanowire was embedded inside the nanopores by electrochemical deposition. After partial etching of the alumina walls from the top, Fe nanowires were revealed and then oxidized. Finally, the remaining alumina walls were etched away, and at this stage, all of the Fe nanowires were converted into Fe_2O_3 nanowires by annealing under air. The diameter of a single nanowire is about 50 nm, and the density of nanowires for all observed specimens is about 10^{10} wires/cm^2.

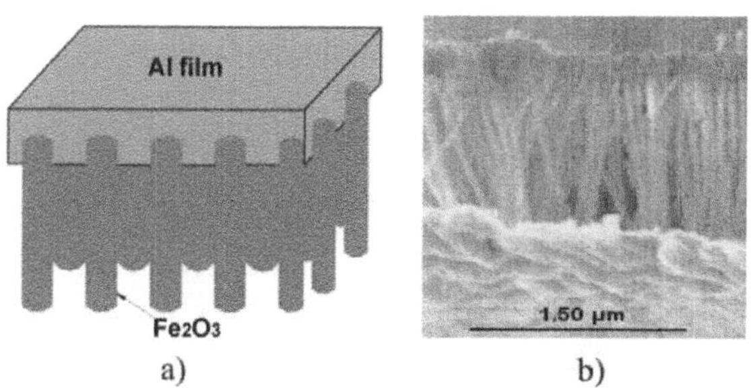

Figure 4.21. *a) Schematic of Al/Fe_2O_3 nanowired systems as proposed by Menon et al. b) SEM image of the nanowired thermite material [MEN 04] (Copyright 2004 American Institute of Physics)*

The main advantages of this approach is the ability to obtain extremely high packing density, intimate contact and precise control between the oxidizer and fuel dimensions. However, the preparation method, which consists of eight steps, is rather complicated. Another considerably simpler technique based on CuO nanowires was later developed by Zhang *et al.* (see Figure 4.22). A Cu film was first deposited onto a silicon substrate either by evaporation or by electro deposition. Cu is then oxidized by thermal annealing in static air and under controlled conditions. The thermal oxidation makes copper oxide nanowires grow. Finally, a thin layer of Al is evaporated around the copper oxide nanowires and on the surface of the underlying CuO/Cu_2O film.

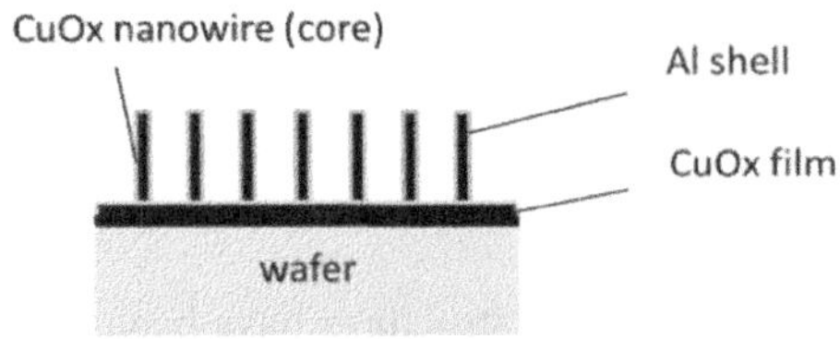

Figure 4.22. *Schematic of nanowired core–shell concept as proposed in [ZHA 07c]*

The advantages of the core–shell structure include enhanced fuel–oxidizer contact, reduced impurity and easy integration into microelectromechanical systems (MEMS) chip. The mixing uniformity and activation energy of Al/CuO_x nanowires were significantly improved and reduced, respectively, compared with those of Al/CuO nanopowder mixtures. However, the formation of CuO nanowires through the thermal annealing of Cu thin films inevitably results in the undesirable generation of a micrometer-scale Cu_2O/CuO layer beneath the CuO nanowires with very similar procedures.

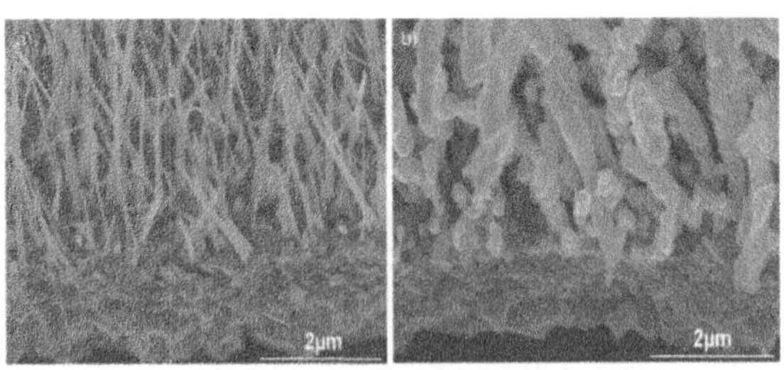

Figure 4.23. *Scanning electron microscopy micrographs of the Al/CuO nanowired core–shell structure a) before Al deposition and b) after Al deposition [ZHA 07c]*

After this first demonstration of the concept, other works have produced core/shell CuO nanowires [OHK 11, PET 10a, YAN 12].

Another core/shell energetic structure has been proposed by Prakash *et al.* in [PRA 05]. They reported a novel aerogel-based process to create a pure nanometer-sized core–shell material obtained by coating a strong oxidizer nanoparticle with a thin layer of relatively mild oxidizer. The authors use a two temperature aerosol spray-pyrolysis method, as shown in Figure 4.23. The core is $KMnO_4$, and the shell is Fe_2O_3. In this process, an aqueous solution of $Fe(NO_3)_3.9H_2O$ and $KMnO_4$ is sprayed into droplets (1 µm in diameter). The environment is dried. The aerosol is then thermally cured. First, it is maintained above the decomposition temperature of iron nitrate (typically 120°C), and then, the temperature is increased to about 240°C (approximately melting point of permanganate). The 0.6 µm composite particles are then collected on a filter. At 120°C, iron nitrate decomposes into Fe_2O_3 enrobing with intimately the solid permanganate. At 240°C, the permanganate melts, and the Fe_2O_3 aggregates around the $KMnO_4$ particles. The resulting composite nanoparticle of $KMnO_4$ (150 nm in diameter) coated with Fe_2O_3 (4 nm thick) has a tunable reactivity. It can be moderated over a relatively large range by changing the thickness of the iron oxide. Besides the resulting intimate contact between the reactants, the oxide coating thickness on each nanoparticle is nanoscale and could be precisely controlled to lower the ignition temperature and also not to inhibit the energetic reaction.

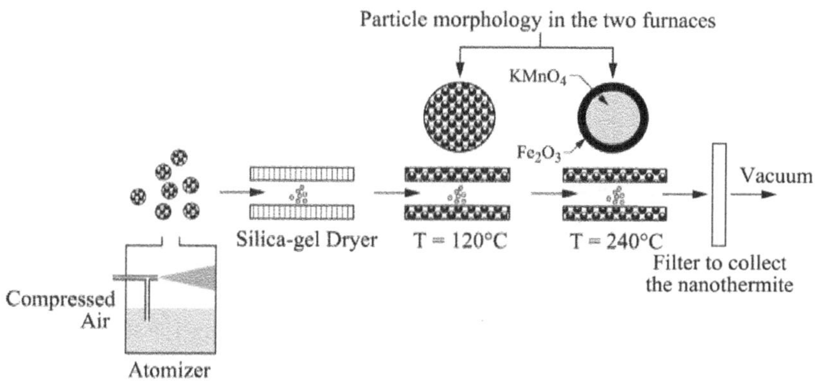

Figure 4.24. *Schematics of the aerosol experimental system for the synthesis of core–shell [PRA 05] (Copyright 2005 American Chemical Society)*

These core–shell energetic materials were mainly characterized by DSC techniques at low heating condition. Most of the studies on nanowires embedded into a metallic matrix experimentally show the decrease in onset temperature (down to 400°C) and an increase in the part of heat of reaction liberated before the melting of Al. Sustained reaction has never been proved since the core–shell structures have not been released from the substrate.

4.5. Reactive porous silicon

Porous silicon has been known as a energetic material when McCord *et al.* [MCC 92] discovered the violent exothermic reaction of porous silicon immersed in nitric acid. The silicon serves as the fuel, and when combined with a source of oxygen, the material becomes highly energetic. In addition to thermal oxidation in air and O_2, it is known that porous silicon is capable of rapid oxidation and possible explosion with liquid O_2, nitric acid, sulfur and strong oxidizing salts. Some experimentations are presented in [KOV 01] and [MIK 02] (see Figure 4.25).

Figure 4.25. *Photos of nanoporous silicon reaction with a $NaClO_4$ oxidizer [CUR 09] (Copyright 2009 IEEE)*

This property placed porous silicon as a possible energetic system for industrial applications as a component of an airbag initiator [HOF 06].

Porous silicon is classically produced by electrochemical etching of bulk silicon in solutions containing fluoride (e.g. HF) and H_2O_2. The porosity and the structure size of the porous silicon layer can be adjusted from 2 to 1,000 nm by selecting suitable etching parameters such as fluoride concentration, current density, treatment duration and starting material. The technique of galvanic corrosion etching has also been developed. In this case, a noble metal such as platinum or gold is directly deposited on Si to serve as a cathode, while the Si acts as the anode to form a galvanic cell that enables Si etching without requiring an external current source. Porosity can reach 83%. The nanopores are then infused with a liquid oxidizer as $Ca(ClO_4)_2$, $KClO_4$, $NaClO_4$ or sulfur. The distribution of nanopores is rather uniform in the porous silicon layer, and the contact between the embedded oxidizers and the silicon is intimate. To enhance the intimacy between the Si and O_2, the porous silicon substrate can also be filled with the oxidizer by vapor deposition method.

Figure 4.26. *SEM images of nanoporous silicon. Left: top view. Right: cross-sectional view [CUR 09] (Copyright 2009 IEEE)*

Plessis [PLE 07] applied a structural model for the porous material to the gravimetric characterization of porosity and pore diameter, proposed a figure of merit for evaluating the strength of the energetic reaction and suggested that the optimum pore diameter for maximum efficiency is around 3.5 nm.

The reaction behaviors of porous-Si/$NaClO_4$ have been systematically studied and results are promising: without the supplementary O_2 environment, the heat of reaction is measured to be ~10 kJ/g, but with supplementary O_2, the reaction yields 27 kJ/g and approaches the theoretical value of 33 kJ/g for complete Si oxidation by $NaClO_4$ which is well above that of zirconium potassium perchlorate (6.3 kJ/g). Becker *et al.* [BEC 11] measured a flame propagation rate of ~3,050 m/s for a porous silicon having pores of 2.4–2.5 nm infused with sodium perchlorate ($NaClO_4$). Moreover, the activation energy of the porous silicon and oxygen reaction can be tailored by controlling the SiO_2 native layer before the oxidizer loading.

In addition to the possible tunability, one significant advantage of reactive porous silicon over the mixtures of fuel and oxidizer nanopowders is its monolithic integration with a Si substrate, making it compatible with microsystems processing techniques.

Currently, important efforts are dedicated to the control reaction dynamics which is an issue.

4.6. Other energetic systems

Electrophoretic deposition (EPD) can be used to deposit thin films with thickness ranging from tens micrometers to several hundreds of micrometers. This technique was proposed to deposit nanothermite films on several kinds of substrates by Sullivan *et al.* from the Lawrence Livermore National Laboratory in 2012 [SUL 12b].

They demonstrated the deposition and patterning of Al/CuO nanothermite film with a density of 29% TMD. They first prepared stable dispersions from nanopowder from a total solids loading of 0.2 vol% in a 3:1 vol% ethanol/water solution. The dispersion was then ultrasonicated for 60 s under 200 W operating at full power and 50% duty cycle. The dispersion was then stirred and sonicated for an additional 60 s, followed by the addition of 10 ml of water [SUL 12b]. The dispersion was stable for 8 h, long enough to realize the deposition done at the highest field strength (200 V/cm).

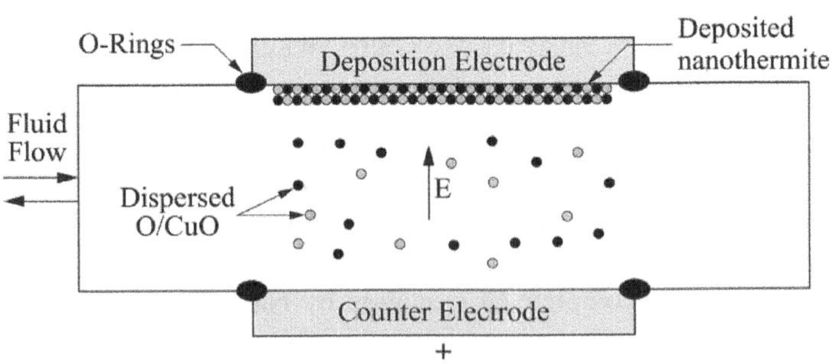

Figure 4.27. *Schematic of EPD cell used by Lawrence Livermore National Laboratory in [SUL 12a] (Copyright 2012 Elsevier)*

An example of a top view and cross-sectional view of Al/CuO nanothermite prepared by EPD is shown in Figure 4.28. The Al/CuO film shows some amount of cracking which occurs during the drying step. The elemental mapping shows the good mixing reached using the EPD technique since no region of significant fuel-oxidizer separation was observed. Optical images of the depositions are shown in Figure 4.29. The equivalence ratio in the as-deposited film was examined as a function of the composition of the precursor dispersion. Different surface charging between Al and CuO can lead to different deposition rates, however, this can be adjusted with a linear correction factor, assuming that the deposition rate scales linearly within the concentration range used. In the work presented in [SUL 12a], Al/CuO film equivalence ratio ($\varnothing_f$) is a linear function of the weighed equivalence ratio in the dispersion ($\varnothing_d$): $\phi_f = 0.566 \times \phi_d$. It should be noted that the equivalence ratio correction factor ($\varnothing_f/\varnothing_d$) is system-dependent, and can be nonlinear if some EPD improves.

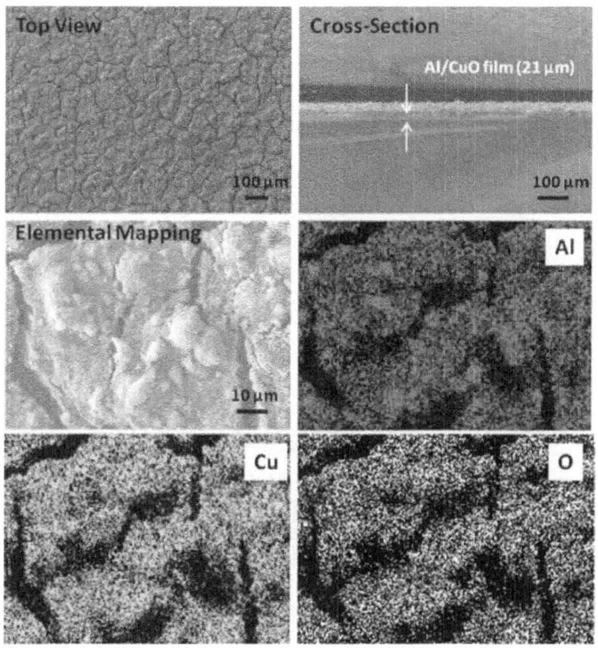

Figure 4.28. *Top and cross-sectional image and elemental mapping of Al/CuO nanothermite obtained by EDP deposition [SUL 12a] (Copyright 2012 Elsevier). For a color version of this figure, see iste.co.uk/rossi/nano.zip*

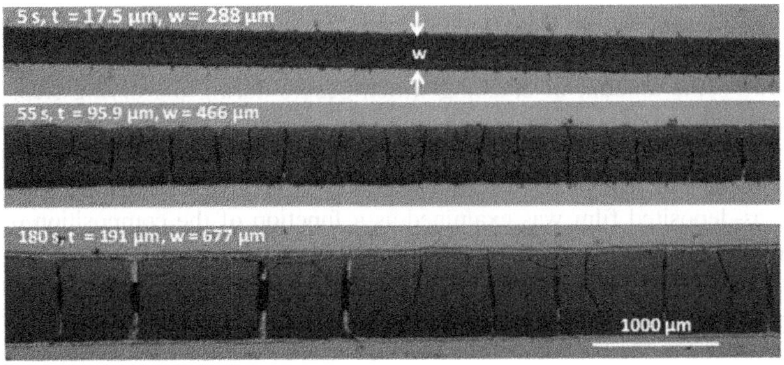

Figure 4.29. *Optical images of some depositions [SUL 12a] (Copyright 2012 Elsevier)*

5

Combustion and Pressure Generation Mechanisms

Considerable research attention has been focused on the combustion and ignition of micrometer-sized and, to a lesser extent, nanosized aluminum particles in oxidizing atmosphere containing oxygen, carbon dioxide and nitrogen with different ratios. Experiments consist of injecting particles of aluminum (Al) in a hydrogen/oxygen/argon microdiffusion flame burner. Even though a pure diffusion model through an oxide shell undergoing phase transitions has been proposed for describing micrometer-sized Al particles combustion [TRU 05], it has been long believed that the reaction of micrometer-sized Al particles occurs in the gas phase similarly to metallic droplet combustion [GLA 64]. However, the combustion of Al nanoparticles is far more complex. The physico-chemical transport mechanisms are still to be unraveled. Some theoretical and experimental studies have focused on nanoparticle oxidation in contact with oxidizing environment [RAI 04] or metallic oxide, [SUL 10]. The interaction between the Al core and its surface passivation oxide, that is characterized by high melting alumina shell, is critical in understanding the ignition mechanism at the nanoscale. Due to the small time and space scales involved in nanothermite mixture, it is difficult to make *in situ* experimental observations to quantitatively validate the potential reaction mechanisms and especially confirm the scenario proposed in this section. Also, a continuum burning regime cannot be reached with metastable nanoparticles as compared to micrometer-sized particles where the mean free path of the gas is significantly less than the particle size. Experimental investigations and models have been proposed in recent years to shed light on nanoparticle combustion mechanisms, and several views have been proposed roughly opposing reactions within the condensed state to gas phase reactions, and diffusion-based models to "mechano-chemical" mechanisms, due to the passivation shell fracturing or even spallation of the shell with projection of Al clusters from the Al core, in all directions.

Figure 5.1 proposes the two main mechanisms for the combustion of an aluminum particle [BAZ 06]. The vapor-phase diffusion or diffusion-limited combustion regime is shown on the left. The flame is detached from the particle surface. Its standoff distance from the particle is regulated by diffusions of both the oxidizer from the atmosphere and the metal from the particle surface. The flame temperature peak is high since it is controlled by the alumina boiling temperature. This structure has been observed for micrometer-sized particles, i.e. with diameter above 10 µm and was proposed in the 1970s and is used in various models [LAW 73].

As the particle diameter is reduced due to the self-sustained combustion process, the diffusion becomes faster than the reaction rates, and the flame moves closer to the surface. We reach a new combustion regime called surface-limited combustion process (middle schematic of Figure 5.1). In this case, the oxygen can diffuse close to the particle surface and engage chemical reactions heterogeneously on the surface. This structure has been observed for particles burning in CO_2 [LEG 01], for instance. In that regime called surface-limited combustion process, the temperature remains near the aluminum boiling temperature. The right schematic corresponds to the shrinking core model proposed for nanoparticles [PAR 05] combustion: the oxidizer diffuses through the alumina shell into the core aluminum region provoking the aluminum/oxygen reaction. Combustion temperatures in this case will be higher in the particle core, site of the reaction. Recent measurements of Al nanoparticles combustion [BAZ 06] appeared to agree well with this model reviewed in more detail in section 5.2.1.

This section summarizes the different ignition and combustion mechanisms published and discussed in the literature. Then a last section is devoted to gas generation during thermite combustion. Major progresses are expected in the near future to elucidate these reaction mechanisms.

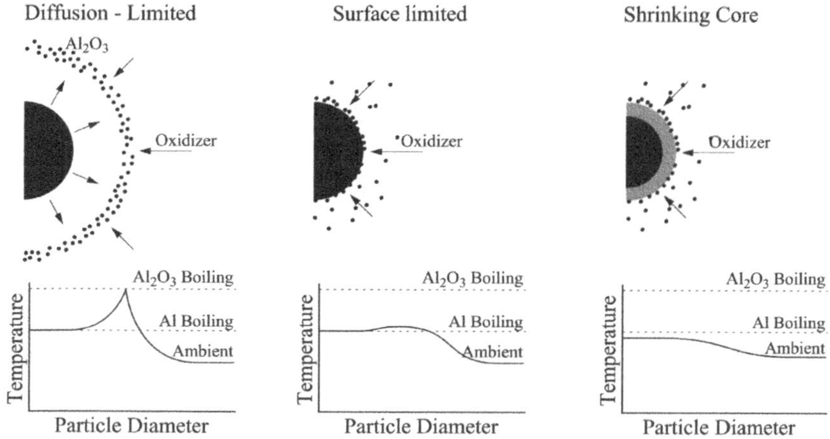

Figure 5.1. *Representation of the different structures for aluminum particle combustion [BAZ 07] (Copyright 2007 Elsevier)*

5.1. General views of Al particle combustion: micro *versus* nano, diffusion-based kinetics

Figure 5.2 shows (i) the experimentally measured aluminum particle combustion rate (r_b) and (ii) ignition temperature as a function of particle diameter [HUA 09, SUN 13]. Data recorded in Figure 5.2(a) concern a system immerged in water. Emphasis is given to the effects of particle size and pressure. According to Sundaram *et al.* [SUN 13], the combustion rate is inversely proportional to the particle diameter. The data plotted in Figure 5.2(b) were obtained under different conditions and particle size distributions; therefore, some caution is required in comparing and analyzing these data. However, the general trends show that for large particles (>100 µm), ignition is achieved at a temperature near the melting point of aluminum oxide, Al_2O_3 (i.e. 2,070 °C). For nanosized Al particles, reaction has been reported to occur at temperatures close to the Al melting point, 660°C.

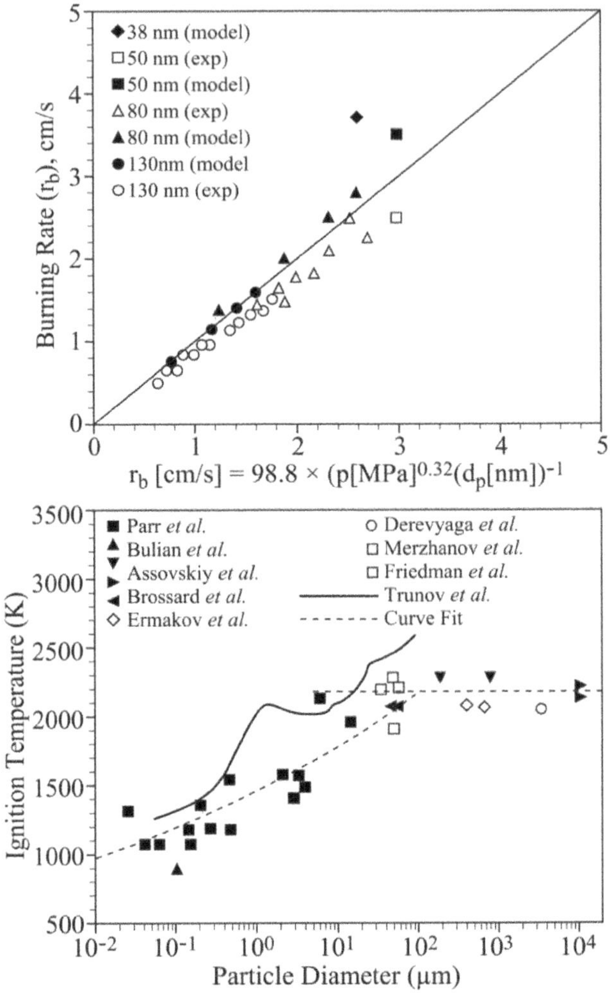

Figure 5.2. *a) Measured and calculated burning rates for different particle diameter and pressures. b) Ignition temperature of aluminum particle as a function of its diameter in O_2 environments [HUA 09] (Copyright 2009 Elsevier)*

These studies have been recently reviewed by Brooks *et al.* [BRO 95]. In this section, we propose a summary of main tendencies and results. In all papers, the particles are assumed to burn completely because elevated temperatures are measured. In addition, there is no reason to expect aerosolidized particles in the gas phase. The burn time (t_b) fits a power law relation with an exponent n:

$$t_b \sim d^n \qquad [5.1]$$

where d is Al particle diameter.

The classical diffusion flame theory yields an exponent of 2, ($t_b \sim d^2$), while the kinetic-limited regime yields an exponent of 1, ($t_b \sim d$). For the combustion of micrometer-sized aluminum particles, the exponent has been determined to be less than 2 ($t_b \sim d^n$ with $n < 2$) certainly because of the presence of the oxide coating. Brooks *et al.* proposed an exponent n equal to 1.8 to fit experimental data. We also note that for particles with a diameter greater than 10 μm, the burn time is mainly a function of the environmental gas temperature and n is reduced to 0.5. In [BAZ 07], Bazyn *et al.* [BRO 95] established that, for micrometer-sized particles, an exponent n, ranging between 0.68 and 1.15, was necessary in order to account for oxidizing atmosphere and burning condition, such as typically the percentage of O_2 and CO_2. In [HUA 07], Huang *et al.* proposed an exponent n equal to 0.3 for nanometer-size particles being ~100 nm in diameter. They also found that the flame temperature can be controlled between 630 and 2,100°C by varying the amount of argon in the oxidizing atmosphere.

In [RAI 06], Rai *et al.* characterized the time of aluminum nanoparticle by the following equation $t \propto d^{1.6\pm0.1}$ (for diameter < 35 nm) where t is the time required for a particle to react and d is the diameter of the particle. This law suggests that the larger particles react faster than the smaller particles if we just consider a shrinking core model. If this process was just a surface reaction in the free molecular regime, we should expect a linear relationship between d and t. If we have diffusion of either only aluminum or only oxygen through the oxide shell and without any pressure gradients, then the power law obtained is $t \propto d^2$ which is in agreement with the shrinking core model [LEV 99] in the absence of pressure gradients due to the melting of the aluminum core.

5.2. Stress in the oxide layer and shrinking core model

Most of the earlier works on the aluminum reaction process stated that the ignition and reaction of aluminum nanoparticles is based on diffusion-based mechanisms where Al and oxide species, ionic or not, diffuse across the alumina oxide shell. However, hot-stage transmission electron microscopy coupled with single particle mass spectrometry [RAI 04] evidenced the mechanical rupture of the oxide coating and saw the melted aluminum flowing out of the nanoparticles. Experiments have been conducted at low and high heating rates. Rai *et al.* [RAI 04] showed experimentally for different heating rates, that the aluminum core melts and exerts a pressure on the oxide shell, causing it to crack.

In other recent works, it was found that at least some of the reaction in thermites occurs via a condensed phase reactive sintering mechanism [SUL 12c, SUL 10]. In this mechanism, it was suggested that the reaction occurs at the fuel-oxidizer interface. As oxidation energy is liberated, it is rapidly conducted away from the interface, and this serves to further heat and melt adjacent particle material. When the material is melted, it can be rapidly delivered to the reacting interface via surface tension/capillary forces. Pressurization of alumina shell is expected to simultaneously occur during metal flowing, via decomposition or sublimation of the oxidizer, and then any unreacted fuel continues to react heterogeneously in the oxidizing environment.

Another series of experiments conducted by Levitas *et al.* in [LEV 07] showed that increasing the furnace temperature from 500 to 800°C, the density of the Al nanoparticles increases from 2.72 to 3.85 g/cm^3. The bulk density of aluminum is 2.7 g/cm^3, and the alumina is ranging from 3 to 3.99 g/cm^3 depending on the polymorph structure. Therefore, this increase in particle density witnesses the oxidation of one part of the core aluminum. Thus, when temperature is increased until 1,100°C, the density decreases which is attributed to the formation of hollow particles. The oxidation progresses by the transport of the oxidizer through the oxide layer to the core, as described in the shrinking core model. In this model, illustrated in Figure 5.2, the surface tension effects cause the alumina coating to strongly adhere to the particle [PAR 05]. The flame moves close to the aluminum particle surface and the maximum temperature is close to the aluminum boiling temperature and near the particle surface. Therefore, oxidizer diffuses to the particle surface and then through the alumina shell. Combustion temperature is high at the Al core, then the oxide layer breaks and Al is flowing out but not necessarily significantly elevated from the ambient.

5.3. Aluminum oxidation through diffusion-reaction mechanisms

In [HUA 09], the ignition behavior of Al/CuO nanothermites was experimented at high heating rates of ~10^5°C/s using a platinum wire. When wire temperature reaches ~1,000°C, the power is shut off and the diffusing species are detected using a fast time-resolved mass spectrometry. The authors find, experimentally, that ignition temperature is well above the melting point of aluminum, and that no self-sustained flame is observed below 977°C. They measured an ignition delay (delay between end of heating and reaction onset) consistent with a diffusion-limited reaction. Ignition delay is experimentally seen to increase with the thickening of alumina shell thickness. For other thermites as Al/Fe_2O_3, experiments using time-resolved mass spectrometry have also measured significant O_2 release, under rapid heating (~10^5°C/s) [HUA 09, ZHO 10] and well below the flame apparition.

A diffusive mechanism model has been proposed describing the transport through the oxide layer, considering the mass transfer with the energy balance to account for the heat release during oxidation. It does not consider the possible thinning or rupture of the oxide shell upon the high pressure gradient when Al core melts. The assumption is that the reaction is based on the oxidation of aluminum nanoparticles, which is a transport-controlled process. Hence, the intrinsic reaction rate is assumed to be infinite, and the transport flux of oxygen and aluminum inside the alumina shell determines the reacting flux. Initially, the flux of oxygen reacting with the aluminum surface is equivalent to the collision rate in the free molecular regime. As the reaction progresses, alumina shell grows and coats the surface of the particle. Thus, both oxygen and aluminum may diffuse through it. The reaction occurs, therefore, inside the alumina shell as illustrated in Figure 5.3. Diffusion coefficients for oxygen and aluminum in alumina have been widely reported in the literature with important variations in their values which make the quantitative application of such model difficult. We can find values of the diffusion coefficient of oxygen in alumina ranging from 10^{-27} to 10^{-9} m^2/s at 1,200°C. We can find values of diffusion coefficients for aluminum into alumina ranging from 1.5×10^{-19} m^2/s to 1.5×10^{-8} m^2/s at 500°C. The energy (heat) release from the aluminum/oxygen reaction inside the alumina shell is also considered. The implementation of this model is mainly limited by the fact that most physical properties and data supporting the model are not yet available nor validated, for example, concentration of oxygen at the outer surface of the particle, solubility, etc.

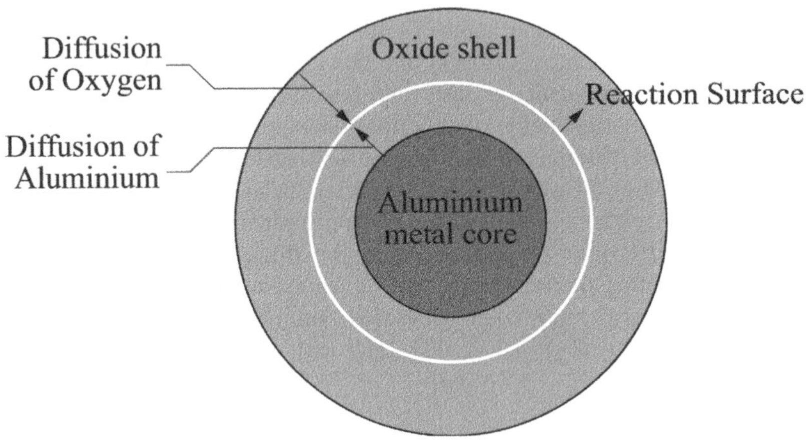

Figure 5.3. *Illustration of the diffusion-reaction mechanism. Schematic of an oxide-coated aluminum particle showing the metal core, oxide shell and the dynamic surface of reaction [RAI 06]*

To include the effect of pressure gradient would require the position of the reaction surface inside the alumina shell which can only be estimated. In the initial stages of reaction, when the coating thickness is small, an increase in pressure gradient causes the oxidation rate to decrease. When the coating thickness is in the range of 1–4 nm, the reaction surface is at a distance of about 0.5 nm from the Al/Al_2O_3 interface. So both aluminum and oxygen diffuse mostly in the region in which convection, due to pressure gradient, acts in the same direction as diffusion, and this enhances the rate of oxidation. Thus, the pressure gradient increases the reaction rate when the thickness of the oxide shell is 1–4 nm. When the coating thickness increases, the reaction zone moves further from the metal/metal oxide interface. Here also, both aluminum and oxygen diffuse in the region in which convection, due to pressure gradient, acts in the same direction as diffusion, and thus increases the reaction rate. This effect gets attenuated when the coating thickness is large because species also have to diffuse through a large region. Thus, we see that at different stages of oxidation the pressure gradient acts in different ways, and this can result in the time–radius relationship.

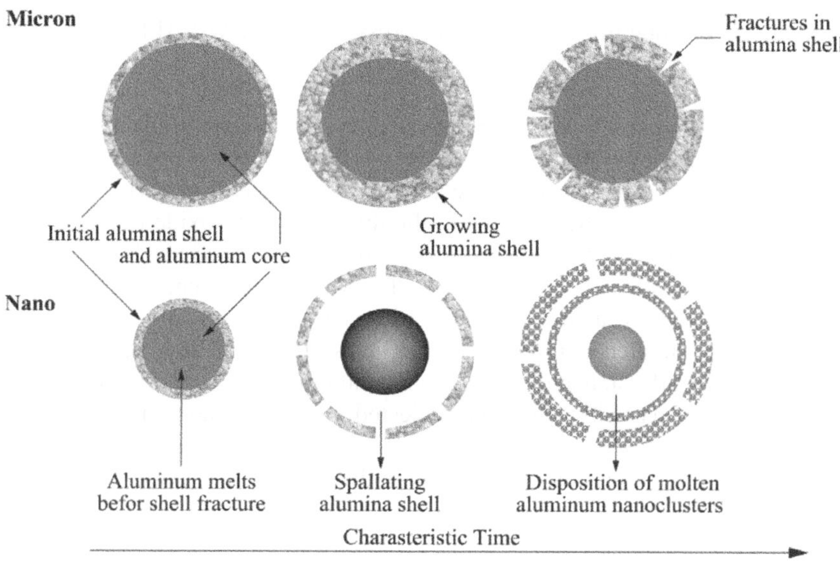

Figure 5.4. *a) Micron-scale particles react by diffusion of aluminum and oxygen through an oxide shell, which fractures before Al melting and then heals; b) nanoscale particles during fast heating react by a melt-dispersion mechanism [LEV 07] (Copyright 2007 American Institute of Physics)*

5.4. Melt-dispersion mechanism

In 2007, Levitas *et al.* brought a series of arguments in favor of a refunding of the modeling view of Al nanoparticle oxidation at elevated temperatures. Among these arguments, the diffusion mechanisms are far too slow to produce flame propagation rates of several hundreds of m/s as often measured in nanothermite mixtures, the flame propagation rate is independent of particle size at the nanoscale (below 80–100 nm), ignition time delay is also found to be independent of particle diameter below 120 nm. This led to the proposition of a new mechanism named the *melt-dispersion mechanism* (MDM) based on the ejection of Al clusters at high velocity from the core to the oxidizer surface after spallation of the oxide shell and subsequent projection of Al nanoclusters into the medium surrounding the particle. In more detail, upon melting of aluminum, the density of aluminum changes from that of solid (2.7 g/cm^3) to a liquid (2.4 g/cm^3) resulting in a high volumic expansion causing the oxide shell to be in tension and the aluminum metal core to be in compression. The large volume change due to melting induces pressures of 1–2 GPa [LEV 07] in the molten Al core and causes dynamic spallation of the shell. Molecular dynamics simulations [CAM 99] have confirmed that an oxide-coated aluminum nanoparticle has substantially high pressure gradients present inside the particle. The aluminum core sustains a positive pressure and the oxide shell is primarily under negative pressure (tension). This pressure gradient inside the particle results in the thinning of the oxide layer until its mechanical rupture. Many Al clusters are, therefore, ejected at a high velocity (estimated around 100–250 m/s). In the MDM approach, the single Al particle covered by a thin alumina shell is transformed into hundreds or thousands of smaller aluminum nanoparticles ejected at high velocity, thus increasing the diffusion and hence, the reaction rate compared to micrometer-size aluminum particle. The molten-ejected Al clusters are spread over the oxidizer surface and can be oxidized in the gas phase (if oxidizing atmosphere) or can partially penetrate into the oxidizer and react.

However, various experimental results for the flame velocity and burn times of nanoparticles are in good qualitative and quantitative agreement with theoretical MDM predictions, as presented in [LEV 07]. In addition to an explanation of the extremely short reaction time, the following correspondence between MDM theory and experiments is obtained: (1) for the particle radius below some critical value, the flame velocity and the ignition time are independent of the radius; (2) damage of the oxide shell suppresses the MDM and promotes the traditional diffusive oxidation mechanism and; (3) nanoflakes do not react as MDM but rather like micron-sized spherical particles.

5.5. Gas and pressure generation in nanothermites

5.5.1. *Thermodynamic models*

The ability of nanothermites to deliver gas species or pressure bursts upon combustion has recently attracted attention, opening routes to a number of novel applications, such as gas delivery [ROD 09, KOR 12, SUL 13, GRI 12, FAN 07, SUN 09] and propulsion systems [APP 09, LAR 03]. Some of the applications will be presented in section 6.3.

A thermite equation giving Al_2O_3 and metal as end reaction products is only a schematic representation of mechanisms occurring during reaction. Indeed, in most practical cases, various intermediates and end products, such as oxides, suboxides and alloys, may be encountered during the reaction process. The formation of gaseous intermediates and their contribution to the pressurization may suggest why a system such as Al/MoO_3, which is thermodynamically predicted to produce approximately half the gas that Al/CuO or Al/Bi2O_3 does, can still react rapidly and exhibit pressures on the same order as the other thermites [SAN 07].

A first attempt to model pressure generation through thermite combustion has been published by Martirosyan *et al.* [WAN 11] based on the equations of gas dynamics. Recently, we proposed the first mechanistic study, i.e. based on identified elementary physico-chemical processes, aimed at predicting the maximum reaction pressure, temperature and reaction products (both gaseous and solid products) of different alumino-thermite mixtures as a function of their theoretical maximum density percentage (%TMD).

A "*local equilibrium thermodynamic*" procedure is employed that allows defining all intermediate compositions, partial and total pressures and temperature as a function of an extent of reaction that follows all defined phase transitions. Practically, considering a stochiometric mixture ($\varnothing = 1$) and assuming that the thermodynamic equilibrium is temporarily reached for each value of the extent of reaction ξ, the overall thermite reaction can be formulated as:

$$\frac{2.b}{3} Al + M_a O_b \rightarrow \frac{b\xi}{3} Al_2O_3 + a.\xi\, M + (1-\xi).\frac{2.b}{3} Al + (1-\xi) M_a O_b \qquad [5.2]$$

where ξ represents the fraction of thermite mixture converted, while the remaining (1-ξ) fraction is still in its initial form.

Additional phases are introduced in equation [5.2] depending on the nature of the oxide, the temperature and the different vapor partial pressures in the reaction chamber. These phases can result from:

– the melting and boiling of all species;

– the decomposition of alloys and oxides;

– the condensation in alloys or oxide forms.

For a given thermite Al/M_aO_b and %TMD, a specific value of the extent of reaction ξ is chosen and the temperature is scanned from ambient to 4,000°C. The partial pressures of various species are then calculated at each temperature and the number of molecules is determined from the perfect gas law. The composition of the condensed phase is then deduced by subtraction. The equation of conservation of energy is applied to calculate the temperatures as a function of extent of reaction ξ. The heat released by the combustion is equal to ξ.%TMD ΔH, with %TMD being the TMD percentage and ΔH, the thermite enthalpy of reaction (see Table 5.1). The energy released is used to increase the temperature of the thermite and to supply the latent heats required for the phase transition to possibly occur. The overall heat quantity *q(T)*, required to heat the system to the desired temperature *T*, can be formulated as:

$$q(T) = \int_{T_0}^{T} [C_v(T) + h(T)].\, dT \qquad [5.3]$$

Boiling species	Latent heat of vaporization (kJ/mol)
Al vaporization	294
Al_2O_3 decomposition	1,402
Cu vaporization	338
CuO decomposition	70
Cu_2O decomposition	115

Table 5.1. *Latent heats of vaporization/decomposition at atmospheric pressure [LID 91]*

where T_0 is the initial temperature, taken here as the ambient temperature (25°C). $C_v(T)$ is the temperature-dependent heat capacity of each species taken from [LID 91] and *h(T)* is the latent heat. The function *h(T)* is composed of a series of Dirac delta functions, at appropriate temperatures, for all phase transitions, except the vaporization considered as a continuous process, in relation to partial vapor pressures.

5.5.2. *Application to Al/CuO*

For Al/CuO, the schematic reaction expressed in equation [5.4] can be broken into nine reactions contained in the six following equations, and taken into account in the simulations.

$$2\ \mathrm{CuO}_{(s)} \leftrightarrow \mathrm{Cu_2O}_{(s)} + \frac{1}{2}\ \mathrm{O}_{2\ (g)}$$

$$\mathrm{Cu_2O}_{(s)} \xrightarrow{1244°C} \mathrm{Cu_2O}_{(l)} \leftrightarrow 2\ Cu_{(l)} + \frac{1}{2}\ \mathrm{O}_{2\ (g)}$$

$$\mathrm{Cu}_{(s)} \xrightarrow{1085°C} \mathrm{Cu}_{(l)} \leftrightarrow 2\ Cu_{(g)}$$

$$\mathrm{Al_2O}_{3\ (s)} \xrightarrow{2072°C} \mathrm{Al_2O}_{3\ (l)} \leftrightarrow 2Al_{(g)} + \frac{3}{2}\ \mathrm{O}_{2\ (g)}$$

$$\mathrm{Al}_{(s)} \xrightarrow{660°C} \mathrm{Al}_{(l)} \leftrightarrow \mathrm{Al}_{(g)} \qquad [5.4]$$

In these equations, the subscripts (s), (l) and (g) stand for solid, liquid and gas phases, respectively. Among these nine reactions, four are phase transitions between condensed phases, represented by direct reactions. A transition temperature is systematically assigned to each of them. The remaining five reactions are reversible (as indicated by the reverse arrows), with the presence of a gas phase. The driving force balancing the direct or reverse reaction is the partial vapor pressure in the gas phase. All thermodynamics constants, including formation energies, have been taken from the literature (see Table 5.1). The Al/CuO equilibrium temperatures as a function of ξ are reported in Figure 5.5. It is observed that the equilibrium temperature increases up to 3,400°C with a slight plateau at 2,072°C corresponding to Al_2O_3 melting. Figure 5.5 shows the maximum pressures for three different %TMD (10, 30 and 50%). The maximum pressure depends not only on the temperature, but also on the volume available for the expansion of the gas. Increasing the %TMD leads to a decrease in the free volume and an increase in the total pressure. For all three %TMD values, the pressure rise starts at $\xi = 0.3$, corresponding to a temperature around 1,000°C (see Figure 5.5).

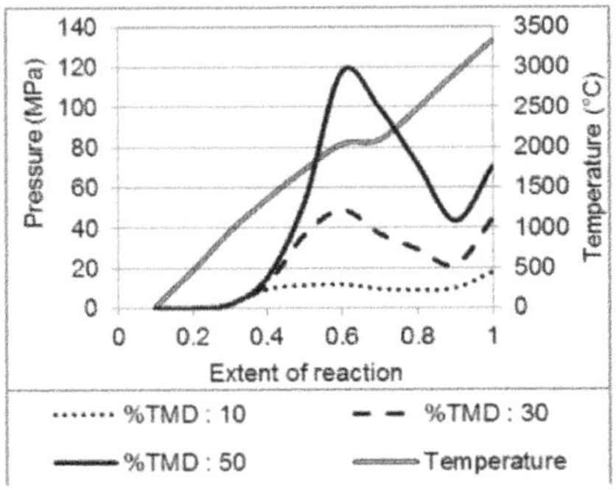

Figure 5.5. *Theoretical pressure and temperature as a function of extent of reaction (ξ) for Al/CuO combustion and for different %TMD*

The pressure rise is not monotonic such that the maximum pressure is reached when the reaction is still incomplete. This result is against the general belief that the pressure drop is only a consequence of reaction stop and subsequent cooling of the system through heat losses. Moreover, for low compaction (TMD percentage of 30%), the pressure evolution clearly shows an oscillating behavior. This trend can also be observed for a higher compaction rate, although the first maximum is less pronounced and rather replaced by a shoulder.

To explain this behavior, Al, Cu and O_2 gas phase partial pressures are given in Figure 5.7, together with the total pressure and the temperature, as a function of ξ and for a TMD percentage of 30%. It shows that all partial pressures do not rise simultaneously.

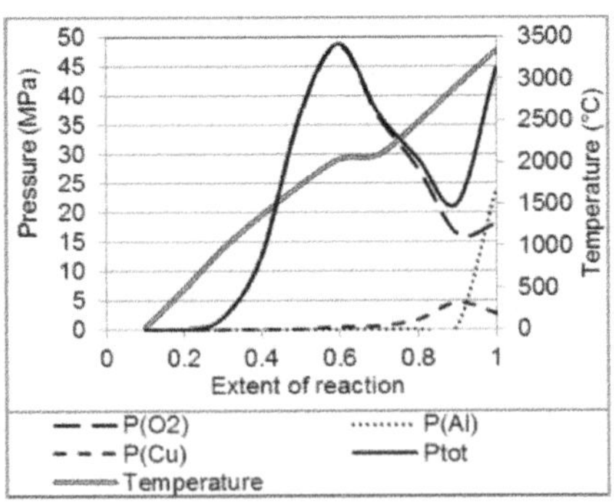

Figure 5.6. *Theoretical partial pressures together with the total pressure and the temperature, as a function of extent of reaction and for TMD percentage of 30%*

The total pressure is mainly driven by the oxygen gas phase which shows a first pressure peak at $\xi = 0.6$ and a gentle pressure shoulder at $\xi = 0.8$, resulting from the decomposition of CuO and Cu_2O, respectively. Aluminum and, to a much lesser extent, copper gas phases only impact the end of the reaction process. The Cu vapor comes from the evaporation of liquid Cu, while Al vapor comes from the decomposition of Al_2O_3. The latter process explains the increase in the total pressure observed from $\xi = 0.9$ to $\xi = 1$. This results from the vaporization of Al_2O_3 due to the high final temperature (3,330°C) that is greater than the boiling temperature of alumina (2,977°C). When the reaction is near completion, the total pressure behavior is almost entirely governed by the aluminum partial pressure. The oxygen partial pressure stabilizes due to compensation between vaporization of the liquid alumina providing oxygen molecules and oxidation reaction with aluminum.

To summarize, for 10% of TMD, the pressure saturation is reached at the end of the reaction ($\xi = 1$) giving rise to a pressure saturation plateau around 18 MPa. For 30 and 50% of TMD, maximum pressures of 49 and 118 MPa are obtained at an extent of reaction $\xi = 0.6$. From these maximum pressure values, the pressure decreases up to $\xi = 0.9$ where the high temperature reached by the system causes the alumina to decompose and in consequence, the pressure to increase again, up to a value of 71 MPa.

6

Applications

In addition to the generation of new primers and new propellant additives, several well-identified applications have already been explored as microignition and rapid initiation, new materials processing, reactive bonding, propulsion, micro-electromechanical systems (MEMS) energy sources, pressure-mediated molecular delivery, material synthesis, biological agent inactivation, hydrogen production and nanochargers for energy storage.

In Table 6.1, expected reaction effects related to these applications are classified.

Used effect	Functionalities and applications
Heat release Radiation	Reactive bonding Microignition and rapid initiation Micropower generation
Gas	Microactuation/propulsion Pressure-mediated molecular delivery
Reaction products	Material synthesis Inactivation of biological agent
Others	Propellant additives Hydrogen production Nanocharges

Table 6.1. *Applications of nanoenergetic materials*

6.1. Reactive bonding

Reactive multilayers can be used to locally produce heat sources for soldering, brazing or welding. In addition, in the case of bimetallic materials, because the total heat released is rather small, reactive multilayered foils are also well suited for low-temperature joining sensitive layers and components with very different coefficients of thermal expansion, e.g. joining metals. The self-propagating exothermic reaction of Al/Ni multilayers was applied in the joining processes of metallic glass [SWI 03], titanium alloy [DUC 04], stainless steel [WAN 04a], silicon wafers [QIU 08] and for MEMS packaging [BRA 12a]. In [SWI 03], the highest shear strength observed between two metallic glasses ($Zr_{57}Ti_5Cu_{20}Ni_8Al_{10}$) is 483 MPa after being bonded by the reaction of 176 µm thick Al/Ni foil under an applied pressure of 160 MPa.

Other works have described the joining of a wide variety of materials including Cu, Au-plated, Al, SiC/Ti and Al/Al2O_3 pairs. Figure 6.1 gives the schematic diagram of the reactive bonding of two components by using reactive foil.

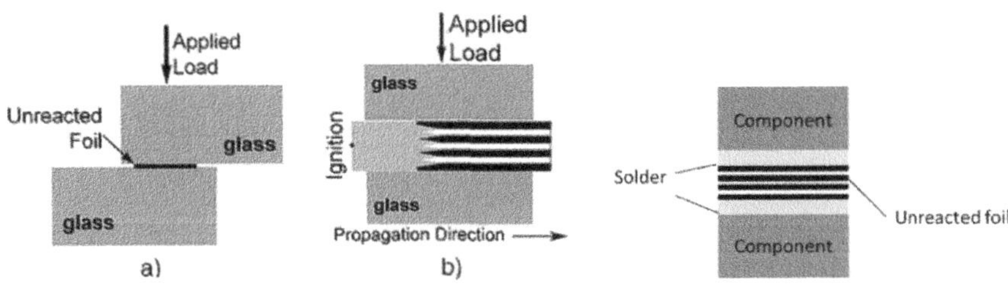

Figure 6.1. *Schematic of the process. Left: without solder a) prior to foil ignition and b) immediately following the ignition. Right: with a solder media [SWI 03] (Copyright 2003 Elsevier)*

The duration of the melting process and the maximum temperature at the solder/component interface generally determine the quality of the final joint. Parameters, such as applied pressure, total foil thickness, layers to be joined and solder selection, can be adjusted to achieve good bonding results. In [WAN 04b], Weihs *et al.* describe the effect of applied pressure (investigated between 48 and 100 MPa) on reactive joining of stainless steel and Al alloy specimens using Al/Ni nanofoils and with AuSn and AgSn solder layers. For a given material system, higher applied joining pressure enhances the flow of the molten solder and thus improves wetting and bonding. The critical applied joining pressure is shown to be dependent on the foil thickness (so, on the total enthalpy of reaction), and the properties of the solder material and components, which determine the duration of melting of the solder and the maximum temperature at the solder/component interface. Longer durations of melting and higher interface temperatures enhance the flow of solder, improve wetting conditions and result in lower critical applied pressures.

In [KOK 09], Kokonou *et al* incorporate bimetallic Al/Ni nanorods on porous alumina template by consecutive evaporation of Al and Ni. The deposition of the materials takes place not only inside the pores but also on the top of the pore walls, resulting in the formation of a porous bimetallic Al/Ni capping layer. In the cross-sectional transmission electron microscope (TEM) image of Figure 6.2, the bimetallic Al/Ni nanorods can be observed inside the pores of the template. Al is at the bottom and Ni is on the top. It is also clearly visible that Al and Ni have been deposited on the surface of the pore walls forming a porous capping layer. Spark ignition experiments were realized on the porous bimetallic films. The reaction propagated radially which produced the melting of nickel aluminides. The melted NiAl was coagulated into microspheres on the porous alumina substrate, proving that the heat of reaction can be served as a nanoscale heat source. This result suggests that reactive Al/Ni nanorods on porous alumina template can provide a thermally actuated valving mechanism for flows into and/or out of anodized aluminum oxide membranes and nanotubes, e.g. for targeted drug delivery.

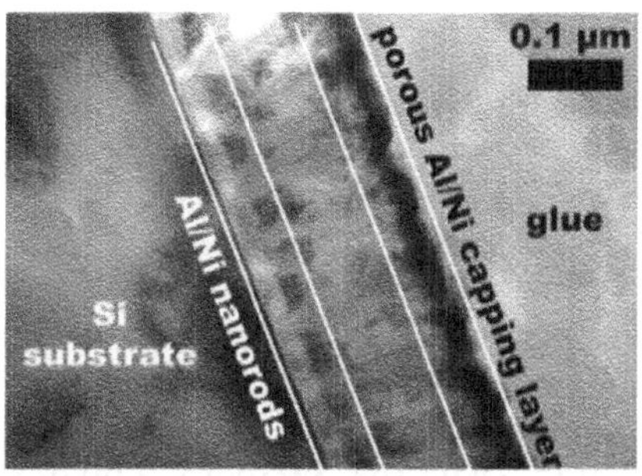

Figure 6.2. *TEM images showing nanorods formed inside the pores of a porous alumina template by the evaporation of 30 nm Al and 30 nm Ni. The porous Al/Ni capping layer can also be seen [KOK 09] (Copyright 2009 Elsevier)*

Al/oxide nanothermites obtained via arrested reactive milling have also been explored for welding applications. In [MOT 12], dense powder made up of 14Al/3CuO/Ni with a decreased ignition temperature was used for welding aluminum alloys. The weld zone thickness in each of the parent alloys was approximately 750 μm. The mean shear strength of the obtained joints has been evaluated at 27 MPa.

6.2. Microignition chips

Nanoenergetic materials have also been explored to produce heat and mechanical work within MEMS devices. This has led to the terminology of nanoenergetic on a chip and pyroMEMS. The idea of pyroMEMS was first proposed by LAAS-CNRS in 1995s: the original concept was the integration of composite solid propellant onto silicon-based devices [ROS 98b] to provide high-power microsources capable of delivering heat or pressure burst for local actuation and microthruster applications. Schematic view and photographs of different kinds of microthrusters are given in Figure 6.2.

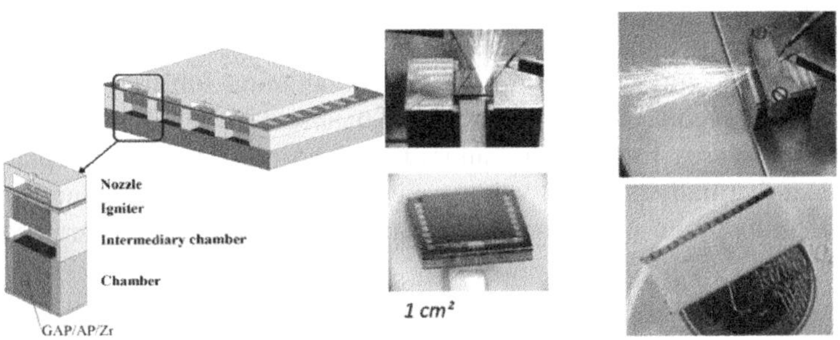

Figure 6.3. *Schematic and photographs of arrays of microthrusters (1 cm × 1 cm)*

Beside the reactive material itself, a fundamental technological challenge associated with harvesting heat and gas produced by a thin layer of reactive material reactions for practical applications entails the ignition within the MEMS devices commonly called the microignition chip [ROS 07, CHU 10a, ZHA 08, QIU 12, TAT 13, CHU 10b, MOR 12, APP 06].

Microignition chips are used in numerous civilian and military applications such as triggering the inflation of air bags in automobiles, propulsion systems, safe-and-arm devices in missiles and rockets, ordnance systems, etc. Traditionally, the technology consists of a metallic hot wire or bridge wire in contact with a highly reactive material in the category of primary or secondary energetic material. The current passing through the wire locally heats the material in contact until its ignition temperature. The first silicon-based pyroMEMS generation was based on electrothermal resistance made on heavily p-doped polysilicon material because of its excellent electric and mechanical properties. Nevertheless, this kind of electrothermal microheater shows some limitations due to its relatively high-power consumption. To improve the microheater performance, and reduce ignition power and energy, the polysicilon resistance was deposited on a few micrometer thick dielectric SiO_2/SiN_x membranes [ROS 98a, ROS 99] (Figure 6.4). However, the use of very thin membrane is poorly reliable for some applications in harsh environments. Glass substrates as support of the electrical resistance were also investigated [ZHA 08] presenting the advantages of robustness and ease in fabrication (see Figure 6.5).

However, glass-based ignitors do not often provide low firing energy (~tens μJ–mJ range) and fast response (difficulty to obtain ignition time < 100 μs).

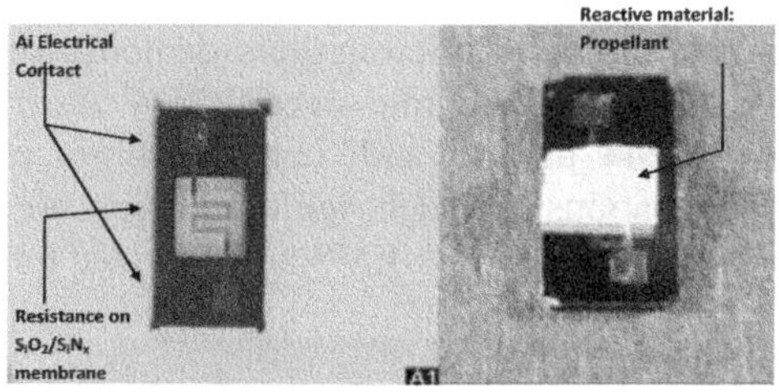

Figure 6.4. *Schematic of the first generation of microignitor made up of polysilicon resistance deposited onto thin dielectric membrane*

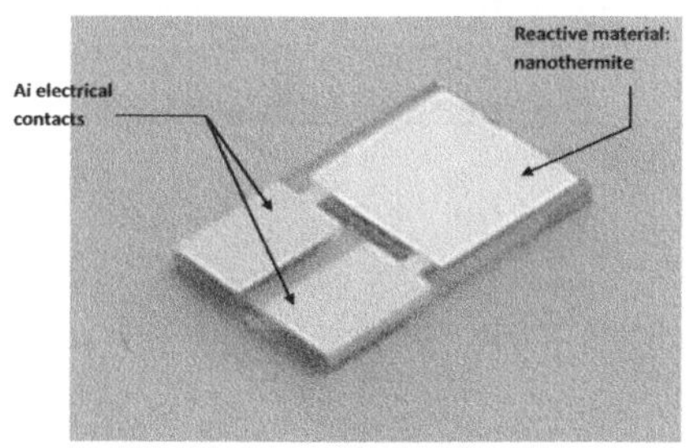

Figure 6.5. *Schematic of the second generation of microignitor made up of metallic resistance deposited onto glass wafer*

In [ZHA 08], Al/CuO nanothermites reported on glass require an ignition energy of 700 μJ, despite heater optimization. In [TAN 08], Tanaka *et al.*

have reported B/Ti reactive igniters on thick silicon membrane. The reported advantages of B/Ti multilayer reactive igniter in comparison with resistive heating igniter are fast responses (well below the milliseconds) and therefore a low firing energy but appear to be not a good solution to ignite propellant because of its too fast response. Recent studies [STA 11d] used silicon bridge wire technology allowing an ignition success rate of 100% over a 30–80 μJ range of firing energies with an ignition time of 2 μs in the high input power operation regime.

Recently, Taton *et al.* [TAT 13] reported the design, realization and characterization of reactive polymeric electrothermal heaters integrating Al/CuO multilayers (see photograph of Figure 6.2) for application in safe arming systems. The reactive Al/CuO multilayered thin foil resides on a 100-μm-thick polyepoxyde/polyethyleneterephtalate (PET) membrane to insulate the reactive layer from the bulk substrate. When current is supplied, Al/CuO reacts and the products of reaction produce sparks and gas. The Al/CuO reactive-material-based microignitors demonstrate the capability to be ignited with a 100% success rate over 0.25–4 A firing current range corresponding to 80–244 μJ and with response times ranging from 2 to 260 μs. Direct applications of such ignition chips are the miniaturized pyrotechnic devices such as fusing MEMS and digitized thrusters.

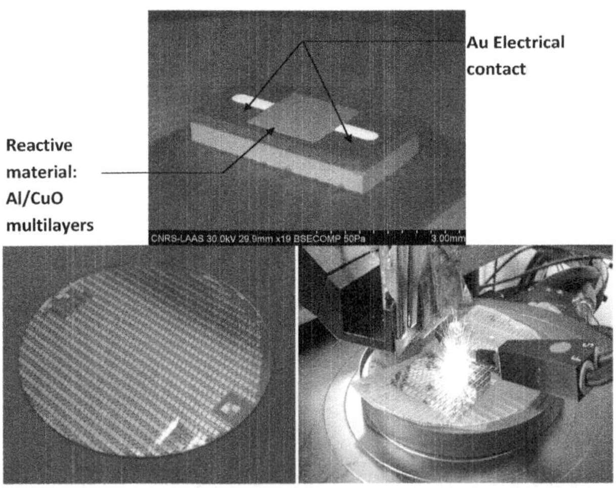

Figure 6.6. *(Top) Scanning Electron Microscopy image of microignitor chip integrating an Al/CuO multilayer foils (1.6 mm × 1.5 mm × 4.5 μm) on thin polymeric membrane. (Bottom) Photographs of one 4 inches wafer with thousands of individual ignitors and photograph of the sparkles produced by the Al/CuO reaction (sample and images from LAAS-CNRS)*

This microignitor chip can be, after designing the resistor and energetic material for the application, directly integrated into air bags, propulsion systems, safe-and-arm devices, ordnance systems, etc.

6.3. Microactuation/propulsion

The production of pressure from pyroMEMS is also of interest to actuate pistons for microfluidic valves, microballoons or to accelerate flyers. It is the fact that the high thermite reaction temperature enables the vaporization of some reaction products such as metal, oxide and reaction intermediates to locally and temporally produce pressure ranging from a few megapascals to hundreds of megapascals as described in previous sections. The pressurization rate ranges from 0.1 to 6 MPa/μs, depending on the nanothermite types and environmental conditions. Indeed, parameters, such as type of oxidizer, compaction and thermal environment, can be adjusted to achieve the expected pressure and pressurization rate depending on the application.

We can distinguish three different categories of actuator: high energy and power actuators, low energy actuators and microthrusters.

6.3.1. *High energetic actuators*

A few studies have considered the dynamic pressures developed by the reactive materials reactions [SUL 13, MAR 11, NEL 13, WU 10] designed for accelerating a thin plastic or metallic foil for fusing application. In that case, a high energy release per unit of volume and a high-pressure peak with the highest pressurization rate is required. Different studies have investigated the burning rate and overpressure generation of different kinds of common nanothermite mixtures prepared using aluminum nanoparticles mixed with different oxidizers (e.g. Bi_2O_3, CuO, MoO_3 and polytetrafluoroethylene (PTFE)) loaded in tiny volumes ranging from 0.009 to 1 cm^3. The largest reported pressure × volume value was 33 Pa × m^3 for a 0.1 g reactant sample.

6.3.2. *Fast impulse nanothermite thrusters*

Al/CuO nanothermite-based reactive nanomaterials have also been developed for microthruster applications in several studies. In [APP 09], the nanothermite-based thrust motors were fabricated using conventional machining processes, typically by boring out stainless steel bolts. The inner diameter of the chamber was approximately 1.59 mm. Different chamber lengths (3.5, 6 and 8.5 mm) were tested with a convergent–divergent nozzle. A schematic of the thruster designs is shown in Figure 6.7.

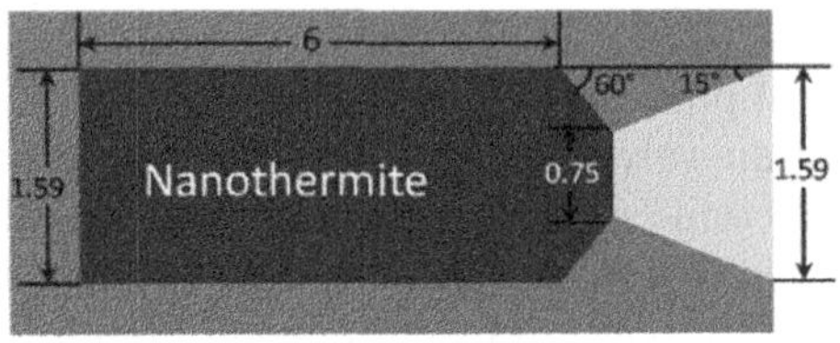

Figure 6.7. *Schematic of thruster design nozzle. All dimensions are in millimeters [APP 09] (Copyright 2009 American Institute of Aeronautics and Astronautics)*

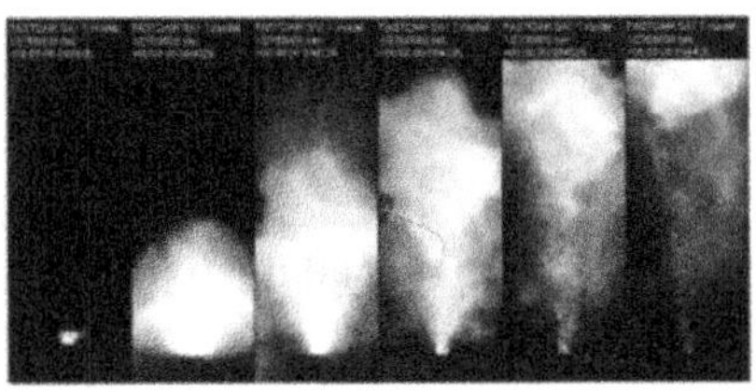

Figure 6.8. *Sequence of high-speed video frames from a thruster tested in the fast regime with the motor without a nozzle [APP 09] (Copyright 2009 American Institute of Aeronautics and Astronautics)*

Various quantities (9–38 mg) of nanothermites are loaded into the chamber by pressing the material over a range of compaction ranging from 20% theoretical maximum density (TMD) to 80% TMD. As the compaction varied, it was found that the material showed two distinct impulse characteristics. At low packing pressure, the combustion was in the fast regime, and the resulting thrust forces were similar to 75 N with duration of less than 50 µs. At high density, the combustion was relatively slow and the thrust forces were 3–5 N with the duration of 1.5–3 ms. In both regimes, the specific impulse generated by the material was 20–25 s.

In the 2000s, Rossi *et al.* proposed MEMS-based thrusters and rockets, the structure of which consists of a sandwich of three micromachined silicon substrates: nozzles, igniters and propellant chambers. Schematic views with typical dimensions are presented in Figure 6.9.

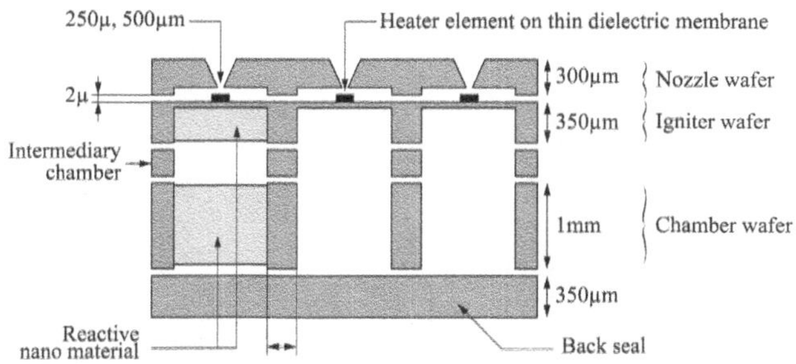

Figure 6.9. *Schematic view of solid propellant microthruster as designed and patented by Rossi et al.*

The nanoenergetic material (being composite propellant) is loaded into the microcavities by screening as presented in Figure 6.7.

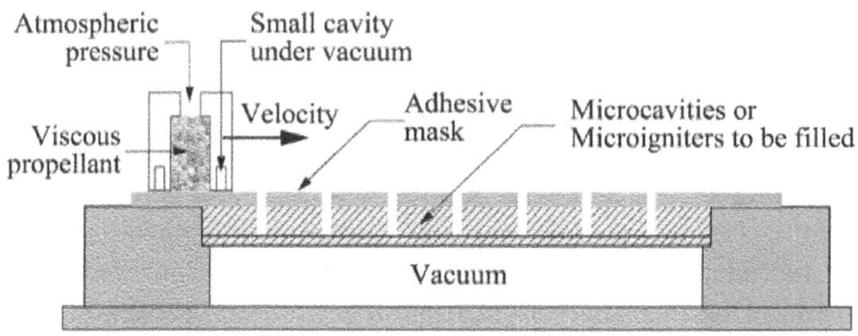

Figure 6.10. *Schematic of the process and the equipment setup developed to load the nanoenergetic material inside thruster microcavities*

Figure 6.11. *Photograph of a fully assembled microthruster chip [ROS 02]*

Figure 6.12. *One photograph during the thrust generation: we clearly see the sparks*

The thrust force issued from the combustion of the solid propellant ranges from 1 mN to a few mN. Specific impulses range from 30 to 50 s depending on the convergent–divergent shape and throat-to-chamber section ratio. Just by geometrical and dimensional considerations, it was possible to adapt the thrust impulse depending on the application requirements. One identified application of those microrockets is space with the realization of microscale thrusters for the control of the attitude or/and the station keeping of very small satellites.

6.3.3. *Smooth actuators*

A low-pressure microactuator consisting of an inflatable polydimethy siloxane (PDMS) elastic membrane and based on the decomposition of a small mass of reactive material deposited on a silicon microstructured platform has been designed and developed in 2009 for disposable lab-on-a-chip applications by Ardila-Rodriguez *et al.* [ROD 09]. The actuator consists of a heating platform made up of a polysilicon resistance on a thin SiO_2/SiN_x dielectric membrane [ROS 99]. A small mass of bimetallic (Mn-Co) reactive powder [SUN 09] is ink-jetted on the heating platform. Then, a thin PDMS membrane covers the nanoenergetic material hermetically. When the actuation is required, the energetic mixture is heated up to 220°C, reacts and generates biocompatible gases mainly composed of N_2, H_2O and O_2, provoking the inflation of the thin PDMS membrane. Figure 6.13 gives three-dimensional and two-dimensional schematic views of the reactive actuator design and structure.

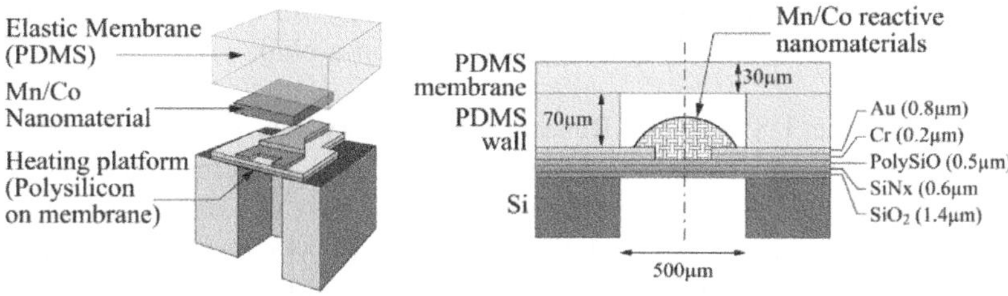

Figure 6.13. *(Left) 3D cut view of the microactuator structure and (right) schematics of the actuator's functioning principle*

Figure 6.14 illustrates the functioning when this miniature actuation platform is integrated into a microchannel for fluid ejection and Figure 6.15 gives photographs of realizations.

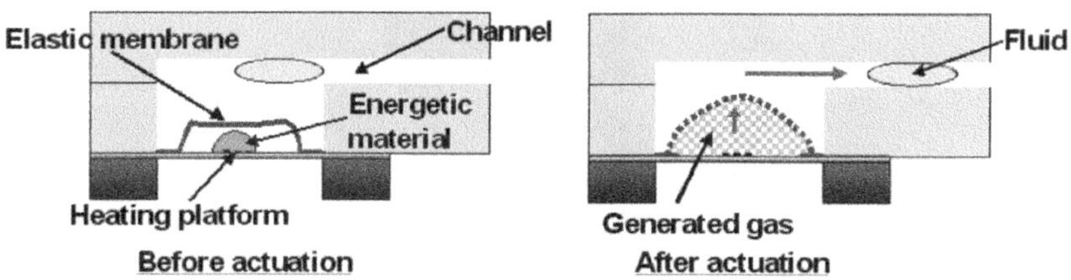

Figure 6.14. *Schematic of the microactuator and dimensional characteristics*

Figure 6.15. *Photographs of realizations: (left) zoom of two actuators and (right) zoom of one actuator*

The characterization of the actuation gave a pressurization of 13 kPa and a membrane deformation of 46 μm for an electrical initial power of 90 mW (6.5 V, 13.9 mA). All these characteristics make such a microactuator well adapted for microfluidic applications and especially for the ejection of fluids contained in microchannels of a disposable lab-on-a-chip.

In 2012, the US Arming Research Laboratory has proposed a similar microfluidic actuator called microfluidic jet injector. The conception follows a vertical design as illustrated in Figure 6.16. Al/CuO nanothermite was selected as reactive material for this study and PDMS membrane was also experimented to transmit the pressure through the fluid. A demonstrator, as shown in Figures 6.17 and 6.18, has been fabricated and assembled from four separately processed Si wafers. Experiments have verified the capacity of the resistive heater to provide controlled and repeatable ignition. At this moment, the filling with reactive nanomaterials and the demonstration of ejection is not revealed.

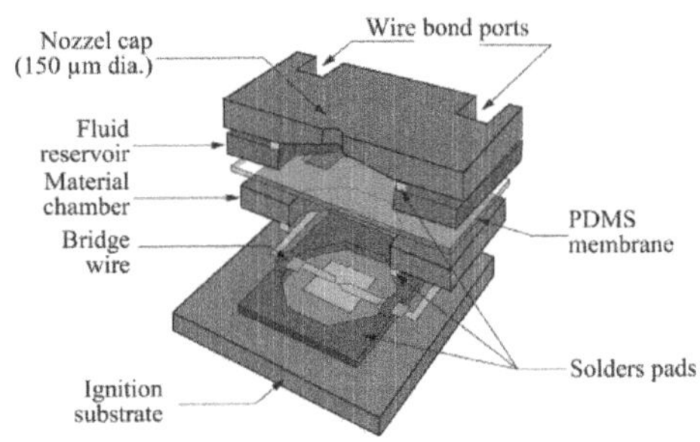

Figure 6.16. *Cross-sectional illustration of microfluidic injector assembly*

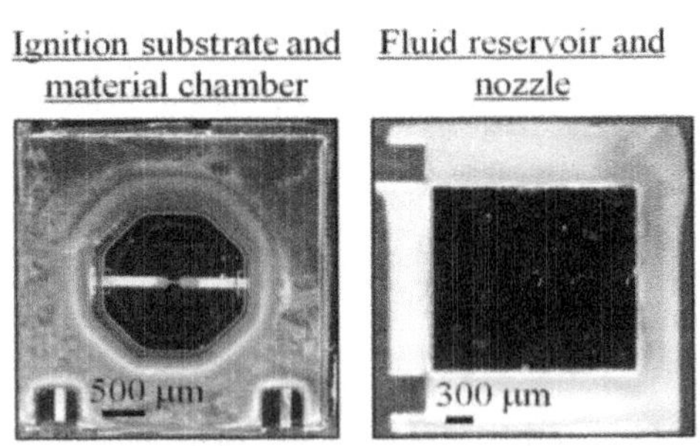

Figure 6.17. *Two-piece partial assembly of a reactive material microfluidic jet injector: ignition substrate bonded with a material chamber (left) and fluid reservoir bonded with a nozzle (right)*

Figure 6.18. *Assembled reactive material microfluid jet injectors: without nozzle (left) and with nozzle (right)*

6.4. Material processing and others

Metal-based energetic nanomaterials have first been explored in synthesis and processing of materials. As already said, thermite reaction between Al and Fe_2O_3 is well known for its use in track welding that forms molten Fe. The explosion of nanoenergetic materials' technological area further stimulates this field of application. Synthesis of intermediary compounds and crystallization of amorphous silicon are two typical examples. The self-sustaining reactions of metallic multilayered foils are exploited to produce intermetallic compounds similarly to the aforementioned reactive bonding. Thermal exploding mode or annealing mode can explain the reaction mechanism depending on the adopted heating rate. Polysilicon is a promising material to fabricate solar cells and thin-film transistors. Crystallization of amorphous silicon with reactive nanomaterials is advantageous for its local heating and swift processing, and is suitable for large-area production.

Power generation using on-chip reactive material has been patented many times, as previously described, and one concept has been patented in [SHU 11]. The basic component is a microinitiation chip on which a thin layer of Al-based reactive materials is deposited, as shown in Figure 6.6. When power is desired, the reactive nanomaterial is ignited and a transducer receives thermal, sonic, magnetic, optic and/or mechanical energy from its combustion and converts it into electrical energy. Preferably, the transducer is a thermoelectric, piezoelectric or magneto device. Preferably, multiple transducers are integrated in one power generator to maximize the power from thermites.

Conclusions

In the last 20 years, research has been very active in synthesis, engineering and characterization of many types of reactive nanomaterials, including metals, bimetallics and mixtures of a metal with many kinds of oxidizers. The development of Al-based energetic nanomaterials as nanothermites eliminated significant limitations of traditional micron-sized metal/oxidizer powder mixtures, including low burning rates and long ignition delays.

Such nanoenergetic materials represent an interesting class of energetic substances most notably because of their high energy densities, adiabatic flame temperature and reaction rates. They have been actively investigated for a wide range of potential applications, including additives in propellants, explosives and pyrotechnics, and more recently also for microignition, mechanical and fluidic actuation, material processing and power generation.

Several synthesis approaches have been investigated and described in this book, including powder mixing, sol–gel processing, self-assembly, vapor deposition and arrested milling. Depending on the preparation method, a broad range of material characteristics and compositions can be obtained with varying particle sizes, surface morphology, interface structure and thickness, and variable densities leading to differences in ignition and combustion mechanisms and overall energetic performances.

Despite a large number of experimentations, as represented in this book, the characterization of ignition and combustion mechanisms remains semi-quantitative. The ignition and combustion of nanoenergetic materials are strongly influenced by the size of constituents and their intimacy: (1) decreasing the particle size generally increases the combustion rate and lowers the ignition temperature and energy; (2) the greater the porosity is, the greater rate the burning is. This trend is the opposite for micro-meter-size material; (3) the burning rates are also a function of stochiometry and compaction and can exhibit different modes of propagation (convective to conductive burning). Nanothermites do not detonate.

Most importantly, the experimental results and data need to be presented in such a way that it will enable us to compare materials prepared using different manufacturing techniques with one another.

In the last decade, advances in the materials characterization techniques, such as high-resolution transmission electron microscopy and nanocalorimetry, have been critically important for progress in the area of nanoenergetic materials and their reaction understanding. In the future, advances in materials characterization techniques are expected to remain important in furthering the understanding of energetic nanomaterial properties and to get insights into the mechanisms occurring at the atomic scale level, especially to understand the role of interfacial layer (layer between the metal and oxidizer). We have seen throughout this book that the quantitative characterizations of the materials morphology, mixing uniformity and interface control are difficult and even impossible.

For example, dense arrested reactive milling (ARM)-prepared nanothermites have passivation layers with a thickness of approximately 0.5 nm, which is much less than the natural oxide passivation layer found in the mixture of nanopowders and also much less than the naturally formed interfacial layer obtained in vapor layer deposition films. The unique interface layer morphology obtained in ARM-prepared nanothermite is thought to cause the exothermic reaction in such materials to begin at lower temperatures compared to other types of nanothermites.

In the future, new experimental approaches to obtain tailored nanoenergetic materials and utilizing the state–of-the-art material characterization equipment need to be developed. Progresses are expected thanks to the extensive use of *in situ* characterization methods.

Some of the most important and poorly understood mechanisms in reactive multilayered nanomaterials or dense ARM-prepared materials occur at the reactive interfaces which play a crucial role in the overall material ignition and its stability at low temperature (before onset temperature). Atomic diffusions through this interface, also called the barrier layer, and any phenomena occurring at the interface (cracking, morphology changes, etc.)

are therefore critical for understanding both the ignition and aging phenomena. Interfacial layer represents a substantial portion of the overall volume of nanoenergetic material and therefore impacts greatly its overall characteristics.

The nature of the reactive interfaces can be as simple as thin layers of amorphous aluminum oxide, but it can also be complex as it is the case in some multilayer structures involving a composition of intermixed atoms that can include individual metals, oxides, suboxides and alloys. Their morphology can be as simple as planar in reactive multilayered foils or much more complex in the case of three-dimensional surfaces in some nanothermite systems.

It is now important to understand what the formation mechanisms for such interfaces are, how these interfacial layers are impacted by the synthesis process, how these layers evaluate under different environmental conditions, how these layers behave at elevated temperatures and how heterogeneous reactions can starting at very low temperatures ultimately lead to ignition in reactive nanomaterials.

A combination of experimental and theoretical efforts will be needed to address these issues. In particular, the development of more sophisticated models is expected to accurately predict formation at such interfaces and their structure and properties. Such predictions will be very important for an in-depth understanding of the reaction mechanisms of the reactive nanomaterials.

Finally, substantial advances are also needed and expected in the ignition and combustion modeling of the reactive nanomaterials. Combustion models are expected to become more consistent between research groups and thereby give more confidence in the predictions of such models.

Bibliography

[ADA 06] ADAMS D.P. *et al.*, "Self-propagating, high-temperature combustion synthesis of rhombohedral AlPt thin films", *Journal of Materials Research*, vol. 21, no. 12, pp. 3168–3179, 2006.

[ADA 15] ADAMS D.P., "Reactive multilayers fabricated by vapor deposition: A critical review", *Thin Solid Films*, vol. 576, pp. 98–128, 2015.

[AND 13] ANDRE B. *et al.*, "High-energy ball milling to enhance the reactivity of aluminum nanopowders", *Materials Letters*, vol. 110, pp. 108–110, 2013.

[APP 06] APPERSON S. *et al.*, "On-chip initiation and burn rate measurements of thermite energetic reactions", *Multifunctional Energetic Materials*, vol. 896, pp. 81–86, 2006.

[APP 07] APPERSON S. *et al.*, "Generation of fast propagating combustion and shock waves with copper oxide/aluminum nanothermite composites", *Applied Physics Letters*, vol. 91, no. 24, 2007.

[APP 09] APPERSON S. *et al.*, "Characterization of nanothermite material for solid-fuel microthruster applications", *Journal of Propulsion and Power*, vol. 25, no. 5, pp. 1086–1091, 2009.

[ARM 03a] ARMSTRONG R.W. *et al.*, "Enhanced propellant combustion with nanoparticles", *Nano Letters*, vol. 3, no. 2, pp. 253–255, 2003.

[ARM 03b] ARMSTRONG R.W., THADHANI N., WILSON W. *et al.* (eds), *Synthesis, Characterization, and Properties of Energetic Reactive Materials*, Materials Research Society, 2003.

[AUM 95] AUMANN C.E., SKOFRONICK G.L., Martin J.A., "Oxidation behavior of aluminum nanopowders", *Journal of Vacuum Science & Technology B*, vol. 13, no. 3, pp. 1178–1183, 1995.

[BAC 13] BACCIOCHINI A. *et al.*, "Reactive structural materials consolidated by cold spray: Al-CuO thermite", *Surface & Coatings Technology*, vol. 226, pp. 60–67, 2013.

[BAD 08] BADGUJAR D.M. *et al.*, "Advances in science and technology of modern energetic materials: an overview", *Journal of Hazardous Materials*, vol. 151, nos. 2–3, pp. 289–305, 2008.

[BAE 10] BAE J.H. *et al.*, "Crystallization of amorphous Si thin films by the reaction of MoO_3/Al nanoengineered thermite", *Thin Solid Films*, vol. 518, no. 22, pp. 6205–6209, 2010.

[BAH 14] BAHRAMI M., TATON G., CONÉDÉRA V. *et al.*, "Magnetron sputtered Al-CuO nanolaminates: effect of stoichiometry and layers thickness on energy release and burning rate", *Propellants, Explosives, Pyrotechnics*, vol. 39, no. 3, pp. 365–373, 2014.

[BAR 96] BARBEE T.W., WEIHS T., Ignitable heterogeneous stratified structure for the propagation of an internal exothermic chemical reaction along an expanding wavefront and method of making the same, US Patent no. 5538795, 23 July 1996.

[BAR 97] BARMAK K., MICHAELSEN C., LUCADAMO G., "Reactive phase formation in sputter-deposited Ni/Al multilayer thin films", *Journal of Materials Research*, vol. 12, no. 1, pp. 133–146, 1997.

[BAR 11] BARAS F., POLITANO O., "Molecular dynamics simulations of nanometric metallic multilayers: reactivity of the Ni-Al system", *Physical Review B*, vol. 84, no. 2, 2011.

[BAT 99] BATTEZZATI L. *et al.*, "Solid state reactions in Al Ni alternate foils induced by cold rolling and annealing", *Acta Materialia*, vol. 47, no. 6, pp. 1901–1914, 1999.

[BAZ 06] BAZYN T., KRIER H., GLUMAC N., "Combustion of nanoaluminum at elevated pressure and temperature behind reflected shock waves", *Combustion and Flame*, vol. 145, no. 4, pp. 703–713, 2006.

[BAZ 07] BAZYN T., KRIER H., GLUMAC N., "Evidence for the transition from the diffusion-limit in aluminum particle combustion", *Proceedings of the Combustion Institute*, vol. 31, pp. 2021–2028, 2007.

[BEC 11] BECKER C.R. *et al.*, "Galvanic porous silicon composites for high-velocity nanoenergetics", *Nano Letters*, vol. 11, no. 2, pp. 803–807, 2011.

[BEG 07] BEGLEY S.M., BREWSTER M.Q., "Radiative properties of MoO_3 and Al nanopowders from light-scattering measurements", *Journal of Heat Transfer-Transactions of the ASME*, vol. 129, no. 5, pp. 624–633, 2007.

[BEL 09] BELONI E., DREIZIN E.L., "Experimental study of ignition of magnesium powder by electrostatic discharge", *Combustion and Flame*, vol. 156, no. 7, pp. 1386–1395, 2009.

[BEL 10] BELONI E., DREIZIN E.L., "Ignition of aluminum powders by electro-static discharge", *Combustion and Flame*, vol. 157, no. 7, pp. 1346–1355, 2010.

[BEN 03] BENKA O., STEINBATZ M., "Oxidation of aluminum studied by secondary electron emission", *Surface Science*, vol. 525, nos. 1–3, pp. 207–214, 2003.

[BES 02] BESNOIN E. *et al.*, "Effect of reactant and product melting on self-propagating reactions in multilayer foils", *Journal of Applied Physics*, vol. 92, no. 9, pp. 5474–5481, 2002.

[BLO 03] BLOBAUM K.J. *et al.*, "Al/Ni formation reactions: characterization of the metastable Al9Ni2 phase and analysis of its formation", *Acta Materialia*, vol. 51, no. 13, pp. 3871–3884, 2003.

[BOC 05] BOCKMON B.S. *et al.*, "Combustion velocities and propagation mechanisms of metastable interstitial composites", *Journal of Applied Physics*, vol. 98, no. 6, 2005.

[BOE 10] BOETTGE B. *et al.*, "Fabrication and characterization of reactive nanoscale multilayer systems for low-temperature bonding in microsystem technology", *Journal of Micromechanics and Microengineering*, vol. 20, no. 6, pp. 1–8, 2010.

[BOI 02] BOIKO V.M., POPLAVSKI S.V., "Self-ignition and ignition of aluminum powders in shock waves", *Shock Waves*, vol. 11, no. 4, pp. 289–295, 2002.

[BRA 12a] BRAEUER J. *et al.*, "A novel technique for MEMS packaging: reactive bonding with integrated material systems", *Sensors and Actuators A – Physical*, vol. 188, pp. 212–219, 2012.

[BRA 12b] BRAEUER J. *et al.*, "Integrated nano scale multilayer systems for reactive bonding in microsystems technology", *Proceedings of the 4th Electronic System-Integration Technology Conference (ESTC)*, 2012.

[BRA 12c] BRAEUER J. *et al.*, "Investigation of different nano scale energetic material systems for reactive wafer bonding", *Semiconductor Wafer Bonding 12: Science, Technology, and Applications*, vol. 50, no. 7, pp. 241–251, 2012.

[BRE 08] BRECHIGNAC C., HOUDY P., LAHMANI M., *Nanomaterials and Nanochemistry*, Springer-Verlag, Berlin, Heidelberg, 2008.

[BRO 95] BROOKS K.P., BECKSTEAD M.W., "Dynamics of aluminum combustion", *Journal of Propulsion and Power*, vol. 11, no. 4, pp. 769–780, 1995.

[BRU 92] BRUNE H. *et al.*, "Surface migration of hot adatoms in the course of dissociative chemisorption of oxygen on Al(111)", *Physical Review Letters*, vol. 68, no. 5, pp. 624–626, 1992.

[BRU 93] BRUNE H. *et al.*, "interaction of oxygen with Al(111) studied by scanning-tunneling-microscopy", *Journal of Chemical Physics*, vol. 99, no. 3, pp. 2128–2148, 1993.

[CAM 99] CAMPBELL T. *et al.*, "Dynamics of oxidation of aluminum nanoclusters using variable charge molecular-dynamics simulations on parallel computers", *Physical Review Letters*, vol. 82, no. 24, pp. 4866–4869, 1999.

[CHA 06] CHAMPION Y., "Evaporation and condensation for metallic nanopowders", *Annales de Chimie-Science des Matériaux*, vol. 31, no. 3, pp. 281–294, 2006.

[CHE 10] CHENG J.L. *et al.*, "Kinetic study of thermal- and impact-initiated reactions in Al-Fe_2O_3 nanothermite", *Combustion and Flame*, vol. 157, no. 12, pp. 2241–2249, 2010.

[CHI 97a] CHIAVERINI M.J., KUO K.K., PERETZ A. *et al.*, "Heat flux and internal ballistic characterization of a hybrid rocket motor analog", *AIAA Paper 97–3080*, July 1997.

[CHI 97b] CHIAVERINI M.J., SERIN N., JOHNSON D.K. *et al.*, "Instantaneous regression behavior or HTPB solid fuels burning with 60x in a simulated hybrid motor", in KUO K.K. *et al.* (eds), *Challenges in Propellants and Combustion: 100 Years After Nobel*, Begell House, New York, pp. 719–733, 1997.

[CHO 10] CHOWDHURY S. *et al.*, "Diffusive vs explosive reaction at the nanoscale", *Journal of Physical Chemistry C*, vol. 114, no. 20, pp. 9191–9195, 2010.

[CHU 10a] CHURAMAN W.A., CURRANO L., BECKER C., "Initiation and reaction tuning of nanoporous energetic silicon", *Journal of Physics and Chemistry of Solids*, vol. 71, no. 2, pp. 69–74, 2010.

[CHU 10b] CHURAMAN W.A. *et al.*, "Optical initiation of nanoporous energetic silicon for safing and arming technologies", *Proceedings of the Optical Technologies for Arming, Safing, Fuzing, and Firing Vi*, vol. 7795, 2010.

[CHU 12] CHURAMAN W.A. *et al.*, "The first launch of an autonomous thrust-driven microrobot using nanoporous energetic silicon", *Journal of Microelectromechanical Systems*, vol. 21, no. 1, pp. 198–205, 2012.

[CIA 04] CIACCHI L.C., PAYNE M.C., "'Hot-atom' O-2 dissociation and oxide nucleation on Al(111)", *Physical Review Letters*, vol. 92, no. 17, 2004.

[CLA 10] CLARK B.R., PANTOYA M.L., "The aluminium and iodine pentoxide reaction for the destruction of spore forming bacteria", *Physical Chemistry Chemical Physics*, vol. 12, no. 39, pp. 12653–12657, 2010.

[CLE 90] CLEVENGER L.A., THOMPSON C.V., TU K.N., "Explosive silicidation in nickel amorphous-silicon multilayer thin-films", *Journal of Applied Physics*, vol. 67, no. 6, pp. 2894–2898, 1990.

[COA 07] COACH: COMPUTER AIDED CHEMISTRY, Saint Martin d'Hères, France: Thermodata, available at: http://thermodata.online.fr/coachang.html, 2007.

[CUR 09] CURRANO L.J., CHURAMAN W.A., "Energetic nanoporous silicon devices", *Journal of Microelectromechanical Systems*, vol. 18, no. 4, pp. 799–807, 2009.

[DEA 13] DEAN S.W. *et al.*, "Energetic intermetallic materials formed by cold spray", *Intermetallics*, vol. 43, pp. 121–130, 2013.

[DIM 89] DIMITRIOU P. *et al.*, "Laser-induced ignition in solid-state combustion", *Aiche Journal*, vol. 35, no. 7, pp. 1085–1096, 1989.

[DLO 06] DLOTT D.D., "Thinking big (and small) about energetic materials", *Materials Science and Technology*, vol. 22, no. 4, pp. 463–473, 2006.

[DOL 89] DOLINSKII Y.L., YAVOROVSKII N.A., "Effect of current on phase-transition in exploding wires", *Zhurnal Tekhnicheskoi Fiziki*, vol. 59, no. 8, pp. 169–171, 1989.

[DRE 07] DREIZIN E.L. *et al.*, "Reactive nanocomposite materials produced by arrested reactive milling", *Theory and Practice of Energetic Materials*, vol. 7, pp. 3–14, 2007.

[DRE 09] DREIZIN E.L., "Metal-based reactive nanomaterials", *Progress in Energy and Combustion Science*, vol. 35, no. 2, pp. 141–167, 2009.

[DUB 07] DUBOIS C. *et al.*, "Polymer-grafted metal nanoparticles for fuel applications", *Journal of Propulsion and Power*, vol. 23, no. 4, pp. 651–658, 2007.

[DUC 04] DUCKHAM A. *et al.*, "Reactive nanostructured foil used as a heat source for joining titanium", *Journal of Applied Physics*, vol. 96, no. 4, pp. 2336–2342, 2004.

[DUP 07] DU PLESSIS M., "Properties of porous silicon nano-explosive devices", *Sensors and Actuators A – Physical*, vol. 135, no. 2, pp. 666–674, 2007.

[DUP 11] DUPIANO P., STAMATIS D., DREIZIN E.L., "Hydrogen production by reacting water with mechanically milled composite aluminum-metal oxide powders", *International Journal of Hydrogen Energy*, vol. 36, no. 8, pp. 4781–4791, 2011.

[DUT 09] DUTRO G.M. *et al.*, "The effect of stoichiometry on the combustion behavior of a nanoscale Al/MoO(3) thermite", *Proceedings of the Combustion Institute*, vol. 32, pp. 1921–1928, 2009.

[DYE 94] DYER T.S., MUNIR Z.A., RUTH V., "The combustion synthesis of multilayer Nial systems", *Scripta Metallurgica et Materialia*, vol. 30, no. 10, pp. 1281–1286, 1994.

[ECK 93] ECKERT J. *et al.*, "Melting behavior of nanocrystalline aluminum powders", *Nanostructured Materials*, vol. 2, no. 4, pp. 407–413, 1993.

[EDE 94] EDELSTEIN A.S. *et al.*, "Intermetallic phase-formation during annealing of Al/Ni multilayers", *Journal of Applied Physics*, vol. 76, no. 12, pp. 7850–7859, 1994.

[EIS 04] EISENREICH N. *et al.*, "On the mechanism of low temperature oxidation for aluminum particles down to the nano-scale", *Propellants Explosives Pyrotechnics*, vol. 29, no. 3, pp. 137–145, 2004.

[ELL 03] ELLISON R.T.H., MOSER M., *39th AIAA/ASME/SAE/ASEE Joint Propulsion Conference and Exhibit,* Huntville, AL, AIAA 2003-4498, 2003.

[FAD 10] FADENBERGER K. *et al.*, "*In situ* observation of rapid reactions in nanoscale Ni-Al multilayer foils using synchrotron radiation", *Applied Physics Letters*, vol. 97, no. 14, 2010.

[FAN 07] FAN M.Q., XU F., SUN L.X., "Studies on hydrogen generation characteristics of hydrolysis of the ball milling Al-based materials in pure water", *International Journal of Hydrogen Energy*, vol. 32, no. 14, pp. 2809–2815, 2007.

[FED 03] FEDOROV A.V., KHARLAMOVA Y.V., "Ignition of an aluminum particle", *Combustion Explosion and Shock Waves*, vol. 39, no. 5, pp. 544–547, 2003.

[FIS 98] FISCHER S.H., GRUBELICH M.C., "Theoretical energy release of thermites, intermetallics, combustible metals", *Proceedings of the 24th International Pyrotechnics Seminar*, Monterey, CA, pp. 1–6, July 1998.

[FOL 05] FOLEY T., JOHNSON C.E., HIGA K.T., "Inhibition of oxide formation on aluminum nanoparticles by transition metal coating", *Chemistry of Materials*, vol. 17, no. 16, pp. 4086–4091, 2005.

[FOL 07] FOLEY T. *et al.*, "Development of nanothermite composites with variable electrostatic discharge ignition thresholds", *Propellants Explosives Pyrotechnics*, vol. 32, no. 6, pp. 431–434, 2007.

[GAC 05] GACHON J.C. *et al.*, "On the mechanism of heterogeneous reaction and phase formation in Ti/Al multilayer nanofilms", *Acta Materialia*, vol. 53, no. 4, pp. 1225–1231, 2005.

[GAN 11] GANGOPADHYAY S., GANGOPADHYAY K., BEZMELNITSYN A. *et al.*, Shock wave and power generation using on-chip nanoenergetic material, US Patent no. 8066831, 2011.

[GAO 07] GAO Y., Method for preparing nanoscale nickel-coated aluminum powder, Chinese Patent no. 101041180 A 20070926, 2007.

[GAS 01a] GASH A.E. *et al.*, "New sol-gel synthetic route to transition and main-group metal oxide aerogels using inorganic salt precursors", *Journal of Non-Crystalline Solids*, vol. 285, nos. 1–3, pp. 22–28, 2001.

[GAS 01b] GASH A.E. *et al.*, "Use of epoxides in the sol-gel synthesis of porous iron(III) oxide monoliths from Fe(III) salts", *Chemistry of Materials*, vol. 13, no. 3, pp. 999–1007, 2001.

[GAV 00] GAVENS A.J. *et al.*, "Effect of intermixing on self-propagating exothermic reactions in Al/Ni nanolaminate foils", *Journal of Applied Physics*, vol. 87, no. 3, pp. 1255–1263, 2000.

[GLA 64] GLASSMAN I., BRZUSTOWSKI T.A., *Spectroscopic Investigation of Metal Combustion, Heterogeneous Combustion*, Astronautics and Aeronautics Series Academic Press, New York, vol. 15, pp. 41–73, 1964.

[GRA 76] GRANQVIST C.G., BUHRMAN R.A., "Ultrafine metal particles", *Journal of Applied Physics*, vol. 47, no. 5, pp. 2200–2219, 1976.

[GRA 04] GRANIER J.J., PANTOYA M.L., "Laser ignition of nanocomposite thermites", *Combustion and Flame*, vol. 138, no. 4, pp. 373–383, 2004.

[GRI 12] GRINSHPUN S.A. *et al.*, "Inactivation of aerosolized bacillus atrophaeus (BG) endospores and MS2 viruses by combustion of reactive materials", *Environmental Science & Technology*, vol. 46, no. 13, pp. 7334–7341, 2012.

[GRO 06a] GROMOV A.A., FORTER-BARTH U., TEIPEL U., "Aluminum nanopowders produced by electrical explosion of wires and passivated by non-inert coatings: characterisation and reactivity with air and water", *Powder Technology*, vol. 164, no. 2, pp. 111–115, 2006.

[GRO 06b] GROMOV A. *et al.*, "Characterization of aluminum powders: II. Aluminum nanopowders passivated by non-inert coatings", *Propellants Explosives Pyrotechnics*, vol. 31, no. 5, pp. 401–409, 2006.

[HAD 10a] HADJIAFXENTI A. *et al.*, "Synthesis of reactive Al/Ni structures by ball milling", *Intermetallics*, vol. 18, no. 11, pp. 2219–2223, 2010.

[HAD 10b] HADJIAFXENTI A. *et al.*, "The influence of structure on thermal behavior of reactive Al-Ni powder mixtures formed by ball milling", *Journal of Alloys and Compounds*, vol. 505, no. 2, pp. 467–471, 2010.

[HEB 04] HEBERT R.J., PEREPEZKO J.H., "Deformation-induced synthesis and structural transformations of metallic multilayers", *Scripta Materialia*, vol. 50, no. 6, pp. 807–812, 2004.

[HEM 13] HEMERYCK A. *et al.*, "Bottom-up modeling of Al/Ni multilayer combustion: effect of intermixing and role of vacancy defects on the ignition process", *Journal of Applied Physics*, vol. 113, no. 20, 2013.

[HG 84] HG B., *Modern Methods of Particle Size Analysis*, Chemical Analysis, vol. 73, Wiley, New York, 1984.

[HIG 01] HIGA K.T, Johnson C., HOLLINS R.A., Preparation of fine aluminum powders by solution methods, US Patent 6,179,899, 2001.

[HOF 06] HOFMANN A., LAUCHT H., KOVALEV D. *et al.*, Explosive composition and its use, US Patent no. 20,050,072,502 A1, 2006.

[HOS 07] HOSSAIN M. *et al.*, "Crystallization of amorphous silicon by self-propagation of nanoengineered thermites", *Journal of Applied Physics*, vol. 101, no. 5, 2007.

[HUA 07] HUANG Y. *et al.*, "Combustion of bimodal nano/micron-sized aluminum particle dust in air", *Proceedings of the Combustion Institute*, vol. 31, pp. 2001–2009, 2007.

[HUA 09] HUANG Y. *et al.*, "Effect of particle size on combustion of aluminum particle dust in air", *Combustion and Flame*, vol. 156, no. 1, pp. 5–13, 2009.

[HUN 04] HUNT E.M., PLANTIER K.B., PANTOYA M.L., "Nano-scale reactants in the self-propagating high-temperature synthesis of nickel aluminide", *Acta Materialia*, vol. 52, no. 11, pp. 3183–3191, 2004.

[ING 04] INGENITO A., BRUNO C., "Using aluminum for space propulsion", *Journal of Propulsion and Power*, vol. 20, no. 6, pp. 1056–1063, 2004.

[IVA 94] IVANOV V.G. *et al.*, "Combustion of mixtures of ultradisperse aluminum and gel-like water", *Combustion Explosion and Shock Waves*, vol. 30, no. 4, pp. 569–570, 1994.

[IVA 00] IVANOV V.G. *et al.*, "Specific features of the reaction between ultrafine aluminum and water in a combustion regime", *Combustion Explosion and Shock Waves*, vol. 36, no. 2, pp. 213–219, 2000.

[IVA 03] IVANOV Y.F. *et al.*, "Productions of ultra-fine powders and their use in high energetic compositions", *Propellants Explosives Pyrotechnics*, vol. 28, no. 6, pp. 319–333, 2003.

[JAC 95] JACOBSEN J. *et al.*, "Electronic-structure, total energies, and STM images of clean and oxygen-covered Al(111)", *Physical Review B*, vol. 52, no. 20, pp. 14954–14962, 1995.

[JAY 98a] JAYARAMAN S. *et al.*, "Modeling self-propagating exothermic reactions in multilayer systems", *Phase Transformations and Systems Driven Far from Equilibrium*, vol. 481, pp. 563–568, 1998.

[JAY 98b] JAYARAMAN S. *et al.*, "A numerical study of unsteady self-propagating reactions in multilayer foils", *Proceedings of the 27th Symposium (International) on Combustion*, vol. 1–2, pp. 2459–2467, 1998.

[JAY 99] JAYARAMAN S. *et al.*, "Numerical predictions of oscillatory combustion in reactive multilayers", *Journal of Applied Physics*, vol. 86, no. 2, pp. 800–809, 1999.

[JIA 98] JIANG W.H., YATSUI K., "Pulsed wire discharge for nanosize powder synthesis", *IEEE Transactions on Plasma Science*, vol. 26, no. 5, pp. 1498–1501, 1998.

[JOH 07] JOHNSON C.E. *et al.*, "Characterization of nanometer- to micron-sized aluminum powders: size distribution from thermogravimetric analysis", *Journal of Propulsion and Power*, vol. 23, no. 4, pp. 669–682, 2007.

[JON 00] JONES D.E.G. *et al.*, "Thermal characterization of passivated nanometer size aluminium powders", *Journal of Thermal Analysis and Calorimetry*, vol. 61, no. 3, pp. 805–818, 2000.

[JON 03] JONES D.E.G. *et al.*, "Hazard characterization of aluminum nanopowder compositions", *Propellants Explosives Pyrotechnics*, vol. 28, no. 3, pp. 120–131, 2003.

[JOU 05a] JOUET R.J., WARREN A.D., MANNION J.D., "Preparation and passivation of aluminum nanoparticles for energetics applications", *Abstracts of Papers of the American Chemical Society*, vol. 230, pp. U2250–U2251, 2005.

[JOU 05b] JOUET R.J. *et al.*, "Surface passivation of bare aluminum nanoparticles using perfluoroalkyl carboxylic acids", *Chemistry of Materials*, vol. 17, no. 11, pp. 2987–2996, 2005.

[JOU 06] JOUET R.J. *et al.*, "Preparation and reactivity analysis of novel perfluoroalkyl coated aluminium nanocomposites", *Materials Science and Technology*, vol. 22, no. 4, pp. 422–429, 2006.

[KAP 12] KAPPAGANTULA K.S. *et al.*, "Tuning energetic material reactivity using surface functionalization of aluminum fuels", *Journal of Physical Chemistry C*, vol. 116, no. 46, pp. 24469–24475, 2012.

[KIE 01] KIEJNA A., LUNDQVIST B.I., "First-principles study of surface and subsurface O structures at Al(111)", *Physical Review B*, vol. 63, no. 8, 2001.

[KIE 02] KIEJNA A., LUNDQVIST B.I., "Stability of oxygen adsorption sites and ultrathin aluminum oxide films on Al(111)", *Surface Science*, vol. 504, nos. 1–3, pp. 1–10, 2002.

[KIM 04] KIM S.H., ZACHARIAH M.R., "Enhancing the rate of energy release from nanoenergetic materials by electrostatically enhanced assembly", *Advanced Materials*, vol. 16, no. 20, pp. 1821–1825, 2004.

[KIM 06] KIM H.Y., CHUNG D.S., HONG S.H., "Intermixing criteria for reaction synthesis of Ni/Al multilayered microfolls", *Scripta Materialia*, vol. 54, no. 9, pp. 1715–1719, 2006.

[KIM 11] KIM J.S. *et al.*, "Direct characterization of phase transformations and morphologies in moving reaction zones in Al/Ni nanolaminates using dynamic transmission electron microscopy", *Acta Materialia*, vol. 59, no. 9, pp. 3571–3580, 2011.

[KLE 05] KLEINER K., *Metal:* fuels of the future, available at: www.mng.org.uk/gh/renewable_energy/metal_NS_article.htm, 2005.

[KNE 09] KNEPPER R. *et al.*, "Effect of varying bilayer spacing distribution on reaction heat and velocity in reactive Al/Ni multilayers", *Journal of Applied Physics*, vol. 105, no. 8, 2009.

[KOK 09] KOKONOU M. *et al.*, "Reactive bimetallic Al/Ni nanostructures for nanoscale heating applications fabricated using a porous alumina template", *Microelectronic Engineering*, vol. 86, nos. 4–6, pp. 836–839, 2009.

[KOR 12] KORAMPALLY M. *et al.*, "Transient pressure mediated intranuclear delivery of FITC-Dextran into chicken cardiomyocytes by MEMS-based nanothermite reaction actuator", *Sensors and Actuators B: Chemical*, vol. 171, pp. 1292–1296, 2012.

[KOV 01] KOVALEV D. *et al.*, "Strong explosive interaction of hydrogenated porous silicon with oxygen at cryogenic temperatures", *Physical Review Letters*, vol. 87, no. 6, 2001.

[KUB 62] KUBASCHEWSKI O., HOPKINS B.E., *Oxidation of Metals and Alloys*, Butterworths & Co., London, pp. 319, 1962.

[KUO 93] KUO K.K. *et al.*, "Preignition dynamics of Rdx-based energetic materials under CO_2-laser heating", *Combustion and Flame*, vol. 95, no. 4, pp. 351–361, 1993.

[KWO 01] KWON Y.S. *et al.*, "Ultra-fine powder by wire explosion method", *Scripta Materialia*, vol. 44, nos. 8–9, pp. 2247–2251, 2001.

[KWO 03a] KWON Y.S. *et al.*, "Passivation process for superfine aluminum powders obtained by electrical explosion of wires", *Applied Surface Science*, vol. 211, nos. 1–4, pp. 57–67, 2003.

[KWO 03b] KWON Y.S. *et al.*, "The mechanism of combustion of superfine aluminum powders", *Combustion and Flame*, vol. 133, no. 4, pp. 385–391, 2003.

[KWO 05] KWON Y.S., ILYIN A.P., NAZARENKO O.B., "Electric explosion of wires in multicomponent reactionary liquid ambiences as method for producing nanopowder of complex composition", *Proceedings of Korus 2005*, pp. 211–213, 2005.

[KWO 13] KWON J. *et al.*, "Interfacial chemistry in Al/CuO reactive nanomaterial and its role in exothermic reaction", *ACS Applied Materials & Interfaces*, vol. 5, no. 3, pp. 605–613, 2013.

[LAN 12] LANTHONY C. *et al.*, "On the early stage of aluminum oxidation: an extraction mechanism via oxygen cooperation", *Journal of Chemical Physics*, vol. 137, no. 9, 2012.

[LAN 14] LANTHONY C. *et al.*, "Elementary surface chemistry during CuO/Al nanolaminate-thermite synthesis: copper and oxygen deposition on aluminum (111) surfaces", *ACS Applied Materials & Interfaces*, vol. 6, no. 17, pp. 15086–15097, 2014.

[LAR 03] LARANGOT B. *et al.*, "Solid propellant micro thrusters for space application", *Houille Blanche-Revue Internationale De L'Eau*, vol. 5, pp. 111–115, 2003.

[LAW 73] LAW C.K., WILLIAMS F.A., "Combustion of magnesium particles in oxygen-inert atmosphere", *Combustion Science and Technology*, vol. 7, no. 5, pp. 197–212, 1973.

[LEE 09] LEE C.H. *et al.*, "Crystallization of amorphous silicon thin films using nanoenergetic intermolecular materials with buffer layers", *Journal of Crystal Growth*, vol. 311, no. 4, pp. 1025–1031, 2009.

[LEG 01] LEGRAND B. *et al.*, "Ignition and combustion of levitated magnesium and aluminum particles in carbon dioxide", *Combustion Science and Technology*, vol. 165, pp. 151–174, 2001.

[LEV 99] LEVENSPIEL O., "Chemical reaction engineering", *Industrial & Engineering Chemistry Research*, vol. 38, no. 11, pp. 4140–4143, 1999.

[LEV 07] LEVITAS V.I. *et al.*, "Mechanochemical mechanism for fast reaction of metastable intermolecular composites based on dispersion of liquid metal", *Journal of Applied Physics*, vol. 101, no. 8, 2007.

[LID 91] LIDE D.R., *Handbook of Chemistry and Physics*, 71st ed., CRC Press, Boca Raton, Florida, 1991.

[MA 90] MA E. *et al.*, "Self-propagating explosive reactions in Al/Ni multilayer thin-films", *Applied Physics Letters*, vol. 57, no. 12, pp. 1262–1264, 1990.

[MAL 08] MALCHI J.Y. *et al.*, "The effect of added Al(2)O(3) on the propagation behavior of an Al/CuO nanoscale thermite", *Combustion Science and Technology*, vol. 80, no. 7, pp. 1278–1294, 2008.

[MAL 09] MALCHI J.Y., FOLEY T.J., YETTER R.A., "Electrostatically self-assembled nanocomposite reactive microspheres", *ACS Applied Materials & Interfaces*, vol. 1, no. 11, pp. 2420–2423, 2009.

[MAN 97] MANN A.B. *et al.*, "Modeling and characterizing the propagation velocity of exothermic reactions in multilayer foils", *Journal of Applied Physics*, vol. 82, no. 3, pp. 1178–1188, 1997.

[MAN 10] MANESH N.A., BASU S., KUMAR R., "Experimental flame speed in multi-layered nano-energetic materials", *Combustion and Flame*, vol. 157, no. 3, pp. 476–480, 2010.

[MAR 06] MARIOTH E., KROEBER H., LOEBBECKE S. *et al.*, "Comparison of nanoparticulate thermite mixtures formed by conventional and supercritical fluid processes", *Proceedings of the 37th International Annual Conference of ICT (Energetic Materials)*, Fraunhofer-Institut fur Chemische Technologie, Karlruhe, 2006.

[MAR 11] MARTIROSYAN K.S., "Nanoenergetic gas-generators: principles and applications", *Journal of Materials Chemistry*, vol. 21, no. 26, pp. 9400–9405, 2011.

[MCC 92] MCCORD P., YAU S.L., BARD A.J., "Chemiluminescence of anodized and etched silicon – evidence for a luminescent siloxene-like layer on porous silicon", *Science*, vol. 257, no. 5066, pp. 68–69, 1992.

[MCD 10] MCDONALD J.P. *et al.*, "Rare-earth transition-metal intermetallic compounds produced via self-propagating, high-temperature synthesis", *Journal of Materials Research*, vol. 25, no. 4, pp. 718–727, 2010.

[MEN 98a] MENCH M.M., YEH C.L., KUO K.K., "Propellant burning rate enhancement and thermal behavior of ultra-fine aluminum powders (Alex)", *Proceedings of the 29th International Annual Conference of ICT*, Karlsruhe, Germany, p. 30/1, 30 June–3 July 1998.

[MEN 98b] MENCH M.M. *et al.*, "Comparison of thermal behavior of regular and ultra-fine aluminum powders (Alex) made from plasma explosion process", *Combustion Science and Technology*, vol. 135, nos. 1–6, pp. 269–292, 1998.

[MEN 04] MENON L. *et al.*, "Ignition studies of Al/Fe2O3 energetic nanocomposites", *Applied Physics Letters*, vol. 84, no. 23, pp. 4735–4737, 2004.

[MIK 02] MIKULEC F.V., KIRTLAND J.D., SAILOR M.J., "Explosive nanocrystalline porous silicon and its use in atomic emission spectroscopy", *Advanced Materials*, vol. 14, no. 1, pp. 38–41, 2002.

[MOO 04] MOORE D.S., SON S.E., ASAY B.W., "Time-resolved spectral emission of deflagrating nano-Al and nano-MoO(3) metastable interstitial composites", *Propellants Explosives Pyrotechnics*, vol. 29, no. 2, pp. 106–111, 2004.

[MOO 07] MOORE K., PANTOYA M.L., SON S.F., "Combustion behaviors resulting from bimodal aluminum size distributions in thermites", *Journal of Propulsion and Power*, vol. 23, no. 1, pp. 181–185, 2007.

[MOR 01] MORDOSKY J.W. *et al.*, "Utilization of nano-sized aluminum particles in RP-1 gel propellants for spray combustion in a rocket engine", *Abstracts of Papers of the American Chemical Society*, vol. 221, pp. U609–U609, 2001.

[MOR 10] MORRIS C.J. *et al.*, "Rapid initiation of reactions in Al/Ni multilayers with nanoscale layering", *Journal of Physics and Chemistry of Solids*, vol. 71, no. 2, pp. 84–89, 2010.

[MOR 11] MORRIS C.J. *et al.*, "Streak spectrograph temperature analysis from electrically exploded Ni/Al nanolaminates", *Thin Solid Films*, vol. 520, no. 5, pp. 1645–1650, 2011.

[MOR 12] MORRIS C.J. *et al.*, "Initiation of nanoporous energetic silicon by optically-triggered, residual stress powered microactuators", *Proceedings of the IEEE 25th International Conference on Micro Electro Mechanical Systems (MEMS)*, 2012.

[MOR 13] MORRIS C.J., WILKINS P.R., MAY C.M., "Streak spectroscopy and velocimetry of electrically exploded Ni/Al laminates", *Journal of Applied Physics*, vol. 113, no. 4, 2013.

[MOT 12] MOTLAGH E.B., KHAKI J.V., SABZEVAR M.H., "Welding of aluminum alloys through thermite like reactions in Al-CuO-Ni system", *Materials Chemistry and Physics*, vol. 133, nos. 2–3, pp. 757–763, 2012.

[NAK 98] NAKAGAWA Y. *et al.*, "Synthesis of TiO_2 and TiN nanosize powders by intense light ion-beam evaporation", *Journal of Materials Science*, vol. 33, no. 2, pp. 529–533, 1998.

[NEL 13] NELLUMS R.R. *et al.*, "Effect of solids loading on resonant mixed Al-Bi_2O_3 nanothermite powders", *Propellants Explosives Pyrotechnics*, vol. 38, no. 5, pp. 605–610, 2013.

[OHK 11] OHKURA Y. *et al.*, "Synthesis and ignition of energetic CuO/Al core/shell nanowires", *Proceedings of the Combustion Institute*, vol. 33, pp. 1909–1915, 2011.

[PAL 96] PALASZEWSKI B., ZAKINI J., "Metallized gelled pellants: oxygen/RP-1/aluminum rocket heat transfer and combustion experiment", *32nd AIAA/ASME/SAE/ASEE Joint Propulsion Conference*, AIAA- 96-2622, NASA TM-107309, Lake Buena Vista, FL, July 1996.

[PAL 98] PALASZEWSKI B., IANOVSKI L.S., CARRICK P., "Propellant technologies: far-reaching benefits for aeronautical and space-vehicle propulsion", *Journal of Propulsion and Power*, vol. 14, no. 5, pp. 641–648, 1998.

[PAL 04] PALASZEWSKI B., J.J., BREISACHER K. *et al.*, *40th AIAA/ASME/SAE/ASEE Joint Propulsion Conference*, AIAA-2004-4191, 2004.

[PAN 05] PANTOYA M.L., GRANIER J.J., "Combustion behavior of highly energetic thermites: nano versus micron composites", *Propellants Explosives Pyrotechnics*, vol. 30, no. 1, pp. 53–62, 2005.

[PAN 09a] PANTOYA M.L., DEAN S.W., "The influence of alumina passivation on nano-Al/Teflon reactions", *Thermochimica Acta*, vol. 493, nos. 1–2, pp. 109–110, 2009.

[PAN 09b] PANTOYA M.L., HUNT E.M., "Nanochargers: energetic materials for energy storage", *Applied Physics Letters*, vol. 95, no. 25, 2009.

[PAR 05] PARK K. *et al.*, "Size-resolved kinetic measurements of aluminum nanoparticle oxidation with single particle mass spectrometry", *Journal of Physical Chemistry B*, vol. 109, no. 15, pp. 7290–7299, 2005.

[PAR 06] PARK K., RAI A., ZACHARIAH M.R., "Characterizing the coating and size-resolved oxidative stability of carbon-coated aluminum nanoparticles by single-particle mass-spectrometry", *Journal of Nanoparticle Research*, vol. 8, nos. 3–4, pp. 455–464, 2006.

[PAT 12] PATRO L.N., HARIHARAN K., "Mechanical milling: an alternative approach for enhancing the conductivity of SnF2", *Materials Letters*, vol. 80, pp. 26–28, 2012.

[PEC 85] PECORA R. (ed.), *Dynamic light Scattering: Applications of Photon Correlation Spectroscopy*, Plenum Press, New York, 1985.

[PER 07a] PERRY W.L. *et al.*, "Energy release characteristics of the nanoscale aluminum-tungsten oxide hydrate metastable intermolecular composite", *Journal of Applied Physics*, vol. 101, no. 6, 2007.

[PER 07b] PERUT C., GOLFIECE M., "New solid propellants", *Proceedings of the 7th International Symposium on Special Topics in Chemical Propulsion*, Kyoto, Japan, 17–21 September 2007.

[PET 10a] PETRANTONI M. *et al.*, "Synthesis process of nanowired Al/CuO thermite", *Journal of Physics and Chemistry of Solids*, vol. 71, no. 2, pp. 80–83, 2010.

[PET 10b] PETRANTONI M. *et al.*, "Multilayered Al/CuO thermite formation by reactive magnetron sputtering: nano versus micro", *Journal of Applied Physics*, vol. 108, no. 8, 2010.

[PIV 04] PIVKINA A. *et al.*, "Nanomaterials for heterogeneous combustion", *Propellants Explosives Pyrotechnics*, vol. 29, no. 1, pp. 39–48, 2004.

[PIV 06] PIVKINA A. *et al.*, "Plasma synthesized nano-aluminum powders – structure, thermal properties and combustion behavior", *Journal of Thermal Analysis and Calorimetry*, vol. 86, no. 3, pp. 733–738, 2006.

[PLA 05] PLANTIER K.B., PANTOYA M.L., GASH A.E., "Combustion wave speeds of nanocomposite Al/Fe2O3: the effects of Fe2O3 particle synthesis technique", *Combustion and Flame*, vol. 140, no. 4, pp. 299–309, 2005.

[PRA 05] PRAKASH A., MCCORMICK A.V., ZACHARIAH M.R., "Tuning the reactivity of energetic nanoparticles by creation of a core-shell nanostructure", *Nano Letters*, vol. 5, no. 7, pp. 1357–1360, 2005.

[PRE 05] PRENTICE D., PANTOYA M.L., CLAPSADDLE B.J., "Effect of nanocomposite synthesis on the combustion performance of a ternary thermite", *Journal of Physical Chemistry B*, vol. 109, no. 43, pp. 20180–20185, 2005.

[PUS 06] PUSZYNSKI J.A., BULIAN C.J., SWIATKIEWICZ J.J., "The effect of nanopowder attributes on reaction mechanism and ignition sensitivity of nanothermites", *Multifunctional Energetic Materials*, vol. 896, pp. 147–158, 2006.

[PUS 07] PUSZYNSKI J.A., BULIAN C.J., SWIATKIEWICZ J.J., "Processing and ignition characteristics of aluminum-bismuth trioxide nanothermite system", *Journal of Propulsion and Power*, vol. 23, no. 4, pp. 698–706, 2007.

[QIU 07] QIU X., WANG J., "Experimental evidence of two-stage formation of Al_3Ni in reactive Ni/Al multilayer foils", *Scripta Materialia*, vol. 56, no. 12, pp. 1055–1058, 2007.

[QIU 08] QIU X., WANG J., "Bonding silicon wafers with reactive multilayer foils", *Sensors and Actuators A – Physical*, vol. 141, no. 2, pp. 476–481, 2008.

[QIU 12] QIU X., *et al.*, "A micro initiator realized by reactive Ni/Al nanolaminates", *Journal of Materials Science–Materials in Electronics*, vol. 23, no. 12, pp. 2140–2144, 2012.

[RAB 07] RABINOVICH O.S. *et al.*, "Conditions for combustion synthesis in nanosized Ni/Al films on a substrate", *Physica B: Condensed Matter*, vol. 392, nos. 1–2, pp. 272–280, 2007.

[RAI 04] RAI A. *et al.*, "Importance of phase change of aluminum in oxidation of aluminum nanoparticles", *Journal of Physical Chemistry B*, vol. 108, no. 39, pp. 14793–14795, 2004.

[RAI 06] RAI A. *et al.*, "Understanding the mechanism of aluminium nanoparticle oxidation", *Combustion Theory and Modelling*, vol. 10, no. 5, pp. 843–859, 2006.

[RAM 05] RAMASWAMY A.L., KASTE P., "A 'nanovision' of the physiochemical phenomena occurring in nanoparticles of aluminum", *Journal of Energetic Materials*, vol. 23, no. 1, pp. 1–25, 2005.

[REE 12] REESE D.A., SON S.F., GROVEN L.J., "Preparation and characterization of energetic crystals with nanoparticle inclusions", *Propellants Explosives Pyrotechnics*, vol. 37, no. 6, pp. 635–638, 2012.

[REI 99] REISS M.E. *et al.*, "Self-propagating formation reactions in Nb/Si multilayers", *Materials Science and Engineering A – Structural Materials Properties Microstructure and Processing*, vol. 261, nos. 1–2, pp. 217–222, 1999.

[RIS 03a] RISHA G.A., EVANS B.J., BOYER E. *et al.* (eds), "Nanosized aluminum and boron-based solid fuel characterization in hybrid rocket engine", *Paper AIAA 2003–4593 39th AIAa/ASME/SAE/ASEE Joint Propulsion Conference and Exhibit*, Huntsville, Alabama, 23 July 2003.

[RIS 03b] RISHA G.A., EVANS B.J., BOYER E. *et al.*, *Proceedings of the 39th AIAA/ASME/SAE/ASEE Joint Propulsion Conference and Exhibit*, Huntsville, Alabama, AIAA 2003-4593, 20–23 July 2003.

[RIS 07] RISHA G.A. *et al.*, "Combustion of nano-aluminum and liquid water", *Proceedings of the Combustion Institute*, vol. 31, pp. 2029–2036, 2007.

[ROD 09] RODRIGUEZ G.A.A. *et al.*, "A microactuator based on the decomposition of an energetic material for disposable lab-on-chip applications: fabrication and test", *Journal of Micromechanics and Microengineering*, vol. 19, no. 1, 2009.

[ROG 08] ROGACHEV A.S., "Exothermic reaction waves in multilayer nanofilms", *Uspekhi Khimii*, vol. 77, no. 1, pp. 22–38, 2008.

[ROG 10] ROGACHEV A.S., MUKASYAN A.S., "Combustion of heterogeneous nanostructural systems (review)", *Combustion Explosion and Shock Waves*, vol. 46, no. 3, pp. 243–266, 2010.

[ROS 98a] ROSSI C., TEMPLE-BOYER P., ESTEVE D., "Realization and performance of thin SiO_2/SiNx membrane for microheater applications", *Sensors and Actuators A – Physical*, vol. 64, no. 3, pp. 241–245, 1998.

[ROS 98b] ROSSI C. *et al.*, "Realization, characterization of micro pyrotechnic actuators and FEM modelling of the combustion ignition", *Sensors and Actuators A – Physical*, vol. 70, nos. 1–2, pp. 141–147, 1998.

[ROS 99] ROSSI C., ESTEVE D., MINGUES C., "Pyrotechnic actuator: a new generation of Si integrated actuator", *Sensors and Actuators A – Physical*, vol. 74, nos. 1–3, pp. 211–215, 1999.

[ROS 02] ROSSI C. *et al.*, "Design, fabrication and modeling of solid propellant microrocket-application to micropropulsion", *Sensors and Actuators A–Physical*, vol. 99, nos. 1–2, pp. 125–133, 2002.

[ROS 07] ROSSI C. *et al.*, "Nanoenergetic materials for MEMS: a review", *Journal of Microelectromechanical Systems*, vol. 16, no. 4, pp. 919–931, 2007.

[ROS 08] ROSSI C., "Nano matériaux énergétiques: perspectives d'intégration dans les microsystèmes", *Nanotechnologies, Revue des Techniques de l'ingénieur*, vol. 5050, p. 21, 10 April 2008.

[ROS 10] ROSSI C., ESTEVE A., VASHISHTA P., "Nanoscale energetic materials", *Journal of Physics and Chemistry of Solids*, vol. 71, no. 2, pp. 57–58, 2010.

[ROS 14] ROSSI C., "Two decades of research on nano-energetic materials", *Propellants Explosives Pyrotechnics*, vol. 39, no. 3, pp. 323–327, June 2014.

[ROZ 92] ROZENBAND V.I., VAGANOVA N.I., "A strength model of heterogeneous ignition of metal particles", *Combustion and Flame*, vol. 88, no. 1, pp. 113–118, 1992.

[RUB 02] RUBERTO C., YOURDSHAHYAN Y., LUNDQVIST B.I., "Stability of a flexible polar ionic crystal surface: metastable alumina and one-dimensional surface metallicity", *Physical Review Letters*, vol. 88, no. 22, 2002.

[RUF 07] RUFINO B. *et al.*, "Influence of particles size on thermal properties of aluminium powder", *Acta Materialia*, vol. 55, no. 8, pp. 2815–2827, 2007.

[SAL 10a] SALLOUM M., KNIO O.M., "Simulation of reactive nanolaminates using reduced models: III. Ingredients for a general multidimensional formulation", *Combustion and Flame*, vol. 157, no. 6, pp. 1154–1166, 2010.

[SAL 10b] SALLOUM M., KNIO O.M., "Simulation of reactive nanolaminates using reduced models: II. Normal propagation", *Combustion and Flame*, vol. 157, no. 3, pp. 436–445, 2010.

[SAL 10c] SALLOUM M., KNIO O.M., "Simulation of reactive nanolaminates using reduced models: I. Basic formulation", *Combustion and Flame*, vol. 157, no. 2, pp. 288–295, 2010.

[SAN 07] SANDERS V.E. *et al.*, "Reaction propagation of four nanoscale energetic composites (Al/MoO_3, Al/WO_3, Al/CuO, and Bi_2O_3)", *Journal of Propulsion and Power*, vol. 23, no. 4, pp. 707–714, 2007.

[SAR 07] SARATHI R., SINDHU T.K., CHAKRAVARTHY S.R., "Generation of nano aluminium powder through wire explosion process and its characterization", *Materials Characterization*, vol. 58, no. 2, pp. 148–155, 2007.

[SCH 05] SCHOENITZ M., WARD T.S., DREIZIN E.L., "Fully dense nano-composite energetic powders prepared by arrested reactive milling", *Proceedings of the Combustion Institute*, vol. 30, pp. 2071–2078, 2005.

[SCH 06] SCHEFFLAN R. *et al.*, "Formation of aluminum nanoparticles upon condensation from vapor phase for energetic applications", *Journal of Energetic* Materials, vol. 24, no. 2, pp. 141–156, 2006.

[SED 08] SEDOI V.S., IVANOV Y.F., "Particles and crystallites under electrical explosion of wires", *Nanotechnology*, vol. 19, no. 14, 2008.

[SEV 12] SEVERAC F. *et al.*, "High-energy Al/CuO nanocomposites obtained by DNA-directed assembly", *Advanced Functional Materials*, vol. 22, no. 2, pp. 323–329, 2012.

[SHA 06] SHAFIROVICH E., DIAKOV V., VARMA A., "Combustion of novel chemical mixtures for hydrogen generation", *Combustion and Flame*, vol. 144, nos. 1–2, pp. 415–418, 2006.

[SIE 01] SIEBER H. *et al.*, "Structural evolution and phase formation in cold-rolled aluminum-nickel multilayers", *Acta Materialia*, vol. 49, no. 7, pp. 1139–1151, 2001.

[SIM 97] SIMPSON R.L. *et al.*, "CL-20 performance exceeds that of HMX and its sensitivity is moderate", *Propellants Explosives Pyrotechnics*, vol. 22, no. 5, pp. 249–255, 1997.

[SIM 99] SIMONENKO V.N., ZARKO V.E., "Comparative studying of the combustion behavior of composite propellant containing ultrafine aluminum", *Proceedings of the 30th International Annual Conference of ICT*, Karlsruhe, Germany, p. 21, 29 June–2 July 1999.

[SIP 08] SIPPEL T.R., SON S.F., RISHA G.A. *et al.*, "Combustion and characterization of nanoscale aluminum and ice propellants", *Proceedings of the 44th AIAA/ASME/SAE/ASEE Joint Propulsion Conference and Exhibit (AIAA 2008)*, vol. 5040, 2008.

[SON 07a] SON S.F., YETTER R.A., YANG V., "Introduction: nanoscale composite energetic materials", *Journal of Propulsion and Power*, vol. 23, no. 4, pp. 643–644, 2007.

[SON 07b] Son S.F. *et al.*, "Combustion of nanoscale Al/MoO3 thermite in microchannels", *Journal of Propulsion and Power*, vol. 23, no. 4, pp. 715–721, 2007.

[STA 10] STAMATIS D. *et al.*, "Aluminum burn rate modifiers based on reactive nanocomposite powders", *Propellants Explosives Pyrotechnics*, vol. 35, no. 3, pp. 260–267, 2010.

[STA 11a] STAMATIS D., DREIZIN E.L., "Thermal initiation of consolidated nanocomposite thermites", *Combustion and Flame*, vol. 158, no. 8, pp. 1631–1637, 2011.

[STA 11b] STAMATIS D., DREIZIN E.L., HIGA K., "Thermal initiation of Al-MoO_3 nanocomposite materials prepared by different methods", *Journal of Propulsion and Power*, vol. 27, no. 5, pp. 1079–1087, 2011.

[STA 11c] STAMATIS D. *et al.*, "Consolidation and mechanical properties of reactive nanocomposite powders", *Powder Technology*, vol. 208, no. 3, pp. 637–642, 2011.

[STA 11d] STALEY C.S. *et al.*, "Silicon-based bridge wire micro-chip initiators for bismuth oxide-aluminum nanothermite", *Journal of Micromechanics and Microengineering*, vol. 21, no. 11, 2011.

[STO 13] STOVER A.K. *et al.*, "An analysis of the microstructure and properties of cold-rolled Ni:Al laminate foils", *Journal of Materials Science*, vol. 48, no. 17, pp. 5917–5929, 2013.

[STO 14] STOVER A.K. *et al.*, "Mechanical fabrication of reactive metal laminate powders", *Journal of Materials Science*, vol. 49, no. 17, pp. 5821–5830, 2014.

[STR 93] STRUTT A.J. *et al.*, "Shock synthesis of nickel-aluminides", *Proceedings of the High-Pressure Science and Technology – 1993*, Pts 1 and 2, pp. 1259–1262, 1994.

[SUL 10] SULLIVAN K.T. *et al.*, "*In situ* microscopy of rapidly heated nano-Al and nano-Al/WO3 thermites", *Applied Physics Letters*, vol. 97, no. 13, 2010.

[SUL 12a] SULLIVAN K.T., KUNTZ J.D., GASH A.E., "Electrophoretic deposition and mechanistic studies of nano-Al/CuO thermites", *Journal of Applied Physics*, vol. 112, no. 2, 2012.

[SUL 12b] SULLIVAN K.T. *et al.*, "Electrophoretic deposition of binary energetic composites", *Combustion and Flame*, vol. 159, no. 6, pp. 2210–2218, 2012.

[SUL 12c] SULLIVAN K.T. *et al.*, "Reactive sintering: an important component in the combustion of nanocomposite thermites", *Combustion and Flame*, vol. 159, no. 1, pp. 2–15, 2012.

[SUL 13] SULLIVAN K.T. *et al.*, "Synthesis and reactivity of nano-Ag2O as an oxidizer for energetic systems yielding antimicrobial products", *Combustion and Flame*, vol. 160, no. 2, pp. 438–446, 2013.

[SUN 06] SUN J., PANTOYA M.L., SIMON S.L., "Dependence of size and size distribution on reactivity of aluminum nanoparticles in reactions with oxygen and MoO_3", *Thermochimica Acta*, vol. 444, no. 2, pp. 117–127, 2006.

[SUN 07] SUN J., SIMON S.L., "The melting behavior of aluminum nanoparticles", *Thermochimica Acta*, vol. 463, nos. 1–2, pp. 32–40, 2007.

[SUN 09] SUHARD S. *et al.*, "When energetic materials, PDMS-based elastomers, and microelectronic processes work together: fabrication of a disposable microactuator", *Chemistry of Materials*, vol. 21, no. 6, pp. 1069–1076, 2009.

[SUN 13] SUNDARAM D.S. *et al.*, "Effects of particle size and pressure on combustion of nano-aluminum particles and liquid water", *Combustion and Flame*, vol. 160, no. 10, pp. 2251–2259, 2013.

[SWI 03] SWISTON A.J., HUFNAGEL T.C., WEIHS T.P., "Joining bulk metallic glass using reactive multilayer foils", *Scripta Materialia*, vol. 48, no. 12, pp. 1575–1580, 2003.

[TAN 08] TANAKA S. *et al.*, "Test of B/Ti multilayer reactive igniters for a micro solid rocket array thruster", *Sensors and Actuators A – Physical*, vol. 144, no. 2, pp. 361–366, 2008.

[TAT 13] TATON G. *et al.*, "Micro-chip initiator realized by integrating Al/CuO multilayer nanothermite on polymeric membrane", *Journal of Micromechanics and Microengineering*, vol. 23, no. 10, 2013.

[TEP 96] TEPPER F., IVANOV G.V., "Activated aluminium as a stored energy source for propellants", *Proceedings of the 4th International Symposium on Special Topics in Chemical Propulsion*, Stockholm, Sweden, pp. 636–644, 27–28 May 1996 .

[TEP 00] TEPPER F., "Nanosize powders produced by electro-explosion of wire and their potential applications", *Powder Metallurgy*, vol. 43, no. 4, pp. 320–322, 2000.

[TIL 01] TILLOTSON T.M. *et al.*, "Nanostructured energetic materials using sol-gel methodologies", *Journal of Non-Crystalline Solids*, vol. 285, nos. 1–3, pp. 338–345, 2001.

[TRE 08] TRENKLE J.C. *et al.*, "Phase transformations during rapid heating of Al/Ni multilayer foils", *Applied Physics Letters*, vol. 93, no. 8, 2008.

[TRE 10] TRENKLE J.C. *et al.*, "Time-resolved x-ray microdiffraction studies of phase transformations during rapidly propagating reactions in Al/Ni and Zr/Ni multilayer foils", *Journal of Applied Physics*, vol. 107, no. 11, 2010.

[TRU 05] TRUNOV M.A. *et al.*, "Effect of polymorphic phase transformations in Al_2O_3 film on oxidation kinetics of aluminum powders", *Combustion and Flame*, vol. 140, no. 4, pp. 310–318, 2005.

[TRU 06] TRUNOV M.A., SCHOENITZ M., DREIZIN E.L., "Effect of polymorphic phase transformations in alumina layer on ignition of aluminium particles", *Combustion Theory and Modelling*, vol. 10, no. 4, pp. 603–623, 2006.

[UMB 06a] UMBRAJKAR S.M., SCHOENITZ M., DREIZIN E.L., "Control of structural refinement and composition in Al-MoO3 nanocomposites prepared by arrested reactive milling", *Propellants Explosives Pyrotechnics*, vol. 31, no. 5, pp. 382–389 2006.

[UMB 06b] UMBRAJKAR S.M., SCHOENITZ M., DREIZIN E.L., "Exothermic reactions in Al-CuO nanocomposites", *Thermochimica Acta*, vol. 451, nos. 1–2, pp. 34–43, 2006.

[UMB 08] UMBRAJKAR S.M. *et al.*, "Aluminum-rich Al-MoO_3 nanocomposite powders prepared by arrested reactive milling", *Journal of Propulsion and Power*, vol. 24, no. 2, pp. 192–198, 2008.

[UME 07] UMEZAWA N. *et al.*, "1,3,5-Trinitro-1,3,5-triazine decomposition and chemisorption on Al(111) surface: first-principles molecular dynamics study", *Journal of Chemical Physics*, vol. 126, no. 23, p. 234702, 2007.

[VEC 94] VECCHIO K.S., YU L.H., MEYERS M.A., "Shock synthesis of silicides I: Experimentation and microstructural evolution", *Acta Metallurgica et Materialia*, vol. 42, no. 3, pp. 701–714, 1994.

[WAL 07] WALTER K.C., PESIRI D.R., WILSON D.E., "Manufacturing and performance of nanometric Al/MoO_3 energetic materials", *Journal of Propulsion and Power*, vol. 23, no. 4, pp. 645–650, 2007.

[WAN 01a] WANG Q. *et al.*, "One-step synthesis of the nanometer particles of gamma-Fe_2O_3 by wire electrical explosion method", *Materials Research Bulletin*, vol. 36, nos. 3–4, pp. 503–509, 2001.

[WAN 01b] WANG Q. *et al.*, "Preparation and characterization of nanocrystalline powders of Cu-Zn alloy by wire electrical explosion method", *Materials Science and Engineering A – Structural Materials: Properties Microstructure and Processing*, vol. 307, nos. 1–2, pp. 190–194, 2001.

[WAN 04a] WANG J. *et al.*, "Joining of stainless-steel specimens with nanostructured Al/Ni foils", *Journal of Applied Physics*, vol. 95, no. 1, pp. 248–256, 2004.

[WAN 04b] WANG J. *et al.*, "Investigating the effect of applied pressure on reactive multilayer foil joining", *Acta Materialia*, vol. 52, no. 18, pp. 5265–5274, 2004.

[WAN 11] WANG L., LUSS D., MARTIROSYAN K.S., "The behavior of nanothermite reaction based on Bi_2O_3/Al", *Journal of Applied Physics*, vol. 110, no. 7, 2011.

[WAN 12] WANG S.X. *et al.*, "An investigation into the fabrication and combustion performance of porous silicon nanoenergetic array chips", *Nanotechnology*, vol. 23, no. 43, pp. 1–7, 2012.

[WAN 13] WANG H.Y. *et al.*, "Electrospray formation of gelled nano-aluminum microspheres with superior reactivity", *ACS Applied Materials & Interfaces*, vol. 5, no. 15, pp. 6797–6801, 2013.

[WEI 05] WEISER V. *et al.*, "Influence of ALEX and other aluminum particles on burning behavior of gelled nitromethane propellants", *Theory and Practice of Energetic Materials*, vol. 6, pp. 838–846, 2005.

[WEI 09] WEISMILLER M.R. *et al.*, "Dependence of flame propagation on pressure and pressurizing gas for an Al/CuO nanoscale thermite", *Proceedings of the Combustion Institute*, vol. 32, pp. 1895–1903, 2009.

[WEI 11a] WEINGARTEN N.S., RICE B.M., "A molecular dynamics study of the role of relative melting temperatures in reactive Ni/Al nanolaminates", *Journal of Physics-Condensed Matter*, vol. 23, no. 27, 2011.

[WEI 11b] WEISMILLER M.R. *et al.*, "Effects of fuel and oxidizer particle dimensions on the propagation of aluminum containing thermites," *Proceedings of the Combustion Institute*, vol. 33, pp. 1989–1996, 2011.

[WEI 13a] WEIR C. *et al.*, "Electrostatic discharge sensitivity and electrical conductivity of composite energetic materials", *Journal of Electrostatics*, vol. 71, no. 1, pp. 77–83, 2013.

[WEI 13b] WEIR C., Pantoya M.L., Daniels M.A., "The role of aluminum particle size in electrostatic ignition sensitivity of composite energetic materials", *Combustion and Flame*, vol. 160, no. 10, pp. 2279–2281, 2013.

[WRO 67] WRONSKI C.R.M., "The size dependence of the melting point of small particles of tin", *British Journal of Applied Physics*, vol. 18, no. 12, p. 1731, 1967.

[WU 10] WU C., LEE D., ZACHARIAH M.R., "Aerosol-based self-assembly of nanoparticles into solid or hollow mesospheres", *Langmuir*, vol. 26, no. 6, pp. 4327–4330, 2010.

[YAN 02] YANG Y., HAMBIR S.A., DLOTT D.D., "Ultrafast vibrational spectroscopy imaging of nanoshock planar propagation", *Shock Waves*, vol. 12, no. 2, pp. 129–136, 2002.

[YAN 03] YANG Y.Q. *et al.*, "Fast spectroscopy of laser-initiated nanoenergetic materials", *Journal of Physical Chemistry B*, vol. 107, no. 19, pp. 4485–4493, 2003.

[YAN 12] YANG Y., XU D.G., ZHANG K.L., "Effect of nanostructures on the exothermic reaction and ignition of Al/CuOx based energetic materials", *Journal of Materials Science*, vol. 47, no. 3, pp. 1296–1305, 2012.

[YAN 14] YANG C. *et al.*, "Fabrication and performance characterization of Al/Ni multilayer energetic films", *Applied Physics A: Materials Science and Processing*, vol. 114, no. 2, pp. 459–464, 2014.

[ZHA 07a] ZHANG K., FAN J.-H., HUANG Y.-H. *et al.*, "Content and activity analysis of aluminum powder in nano-aluminum/PS microcapsules", *Hanneng Cailiao/Chinese Journal of Energetic Materials*, vol. 15, no. 5, pp. 482–484, 2007.

[ZHA 07b] ZHAO S.J., GERMANN T.C., STRACHAN A., "Melting and alloying of Ni/Al nanolaminates induced by shock loading: a molecular dynamics simulation study," *Physical Review B*, vol. 76, no. 10, 2007.

[ZHA 07c] ZHANG K. *et al.*, "Development of a nano-Al/CuO based energetic material on silicon substrate", *Applied Physics Letters*, vol. 91, no. 11, 2007.

[ZHA 08] ZHANG K.L. *et al.*, "A nano initiator realized by integrating Al/CuO-based nanoenergetic materials with a Au/Pt/Cr microheater", *Journal of Microelectromechanical Systems*, vol. 17, no. 4, pp. 832–836, 2008.

[ZHA 13] ZHANG F. *et al.*, "*In-situ* preparation of a porous copper based nano-energetic composite and its electrical ignition properties", *Propellants Explosives Pyrotechnics*, vol. 38, no. 1, pp. 41–47, 2013.

[ZHO 10] ZHOU L. *et al.*, "Time-resolved mass spectrometry of the exothermic reaction between nanoaluminum and metal oxides: the role of oxygen release", *Journal of Physical Chemistry C*, vol. 114, no. 33, pp. 14269–14275, 2010.

[ZHO 11] ZHOU X. *et al.*, "Influence of Al/CuO reactive multilayer films additives on exploding foil initiator", *Journal of Applied Physics*, vol. 110, no. 9, 2011.

[ZHU 03] ZHUKOVSKII Y.F., JACOBS P.W.M., CAUSA M., "On the mechanism of the interaction between oxygen and close-packed single-crystal aluminum surfaces", *Journal of Physics and Chemistry of Solids*, vol. 64, no. 8, pp. 1317–1331, 2003.

[ZHU 11] ZHU P. *et al.*, "Energetic igniters realized by integrating Al/CuO reactive multilayer films with Cr films", *Journal of Applied Physics*, vol. 110, no. 7, 2011.

[ZHU 13] ZHU P. *et al.*, "Characterization of Al/CuO nanoenergetic multilayer films integrated with semiconductor bridge for initiator applications", *Journal of Applied Physics*, vol. 113, no. 18, 2013.

Index

B

C

D, E

M

N, O

T

U, V, W Z

Other titles from

iSTE

in

Nanoscience and Nanotechnology

2015

LEPRINCE-WANG Yamin
Piezoelectric ZnO Nanostructure for Energy Harvesting

2014

BALESTRA Francis
Beyond CMOS Nanodevices 1
Beyond CMOS Nanodevices 2

LACAZE Pierre Camille, LACROIX Jean-Christophe
Non-volatile Memories

LATU-ROMAIN Laurence, OLLIVIER Maelig
Silicon Carbide One-dimensional Nanostructures

NICU Liviu, LEÏCHLÉ Thierry
Micro- and Nanoelectromechanical Biosensors

SENEZ Vincent, THOMY Vincent, DUFOUR Renaud
Nanotechnologies for Synthetic Super Non-wetting Surfaces

TRIPON-CANSELIET Charlotte, CHAZELAS Jean
Nanoscale Microwave Engineering: Optical Control of Nanodevices

2012

BELLEVILLE Marc, CONDEMINE Cyril
Energy Autonomous Micro and Nano Systems

CLAVERIE Alain
Transmission Electron Microscopy in Micro-nanoelectronics

LACAZE PIERRE Camille
Nanotechnologies: Concepts, Processing and Applications

2010

BALESTRA Francis
Nanoscale CMOS / Innovative Materials, Modeling and Characterization

QUERLIOZ Damien, DOLLFUS Philippe
The Wigner Monte Carlo Method for Nanoelectronic Devices: Particle Description of Quantum Transport and Decoherence

TAKADOUM Jamal
Nanomaterials and Surface Engineering

VODA Alina
Micro, Nanosystems and Systems on Chips: Modeling, Control and Estimation

2008

DAVIM J. Paulo, JACKSON J.
Nano and Micromachining

NOUAILLHAT Alain
An Introduction to Nanoscience and Nanotechnology

OUISSE Thierry
Electron Transport in Nanostructures and Mesoscopic Devices

2006

RIGNEAULT Hervé *et al.*
Nanophotonics

Printed and bound by CPI Group (UK) Ltd, Croydon, CR0 4YY
13/09/2026
14961079-0006